# ON STONEHENGE

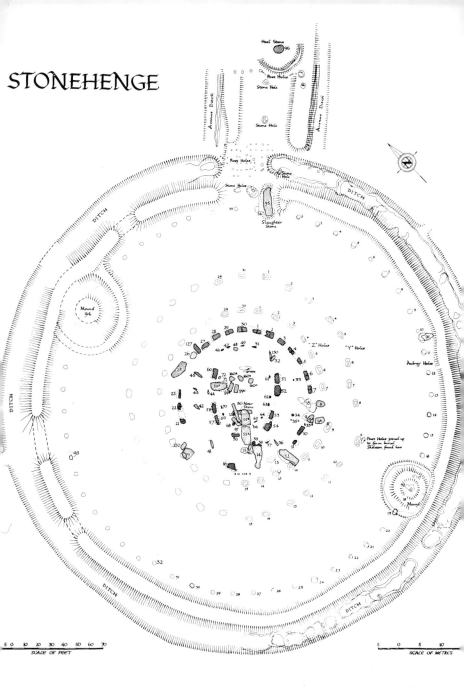

# STONEHENGE

The post holes referred to in the text as A1, A2, A3 and A4 are to be found immediately west of the Heelstone. (Courtesy of the Controller, Her Majesty's Stationery Office. Crown Copyright.)

Fred Hoyle

# ON STONEHENGE

HEINEMANN EDUCATIONAL BOOKS
LONDON

Heinemann Educational Books Ltd

LONDON    EDINBURGH    MELBOURNE    AUCKLAND    TORONTO
HONG KONG    SINGAPORE    KUALA LUMPUR    NEW DELHI
NAIROBI    JOHANNESBURG    LUSAKA    IBADAN
KINGSTON

ISBN 0 435 32958 8
© Fred Hoyle 1977
First published 1977

Published by Heinemann Educational Books Ltd
48 Charles Street    London W1X 8AH
Printed by Butler & Tanner Ltd, Frome and London

# Contents

# Preface

*Hundreds* of thousands of people visit Stonehenge each year. They seem mostly to spend an hour or so walking about the site, wondering what it was all about, as I did on my first visit with my family some twenty-five years ago. The remarkable story discussed and developed in this book goes I believe beyond anything the casual visitor might guess, for it requires the men of the New Stone Age, men living 5,000 years ago, to have been meticulous observers of the night sky, to have calculated with numbers, and to have communicated sophisticated astronomical knowledge among themselves from generation to generation. Since this assessment of the intellectual capacity of neolithic man much exceeds that which archaeologists and prehistorians have hitherto accorded to him, I have felt it important to review the case on which it rests with care, not turning aside from quite complex details, although the reader has been spared the ultimate mathematical forms, at any rate until the appendix at the end of the book.

I wish to thank my wife Barbara for help in assembling the necessary materials for the book, and to acknowledge valuable comments from Professor R. J. C. Atkinson. As always, I am indebted to Jan Rasmussen for her typing of a not too legible manuscript.

*Dockray, Cumbria*                                                        *Fred Hoyle*
*11 February, 1977*

# 1
# Stonehenge

*Stonehenge* lies about two miles west of the town of Amesbury in Wiltshire, close to the V-shaped junction of the roads A303 and A344. Approaching by car from Amesbury, the traveller first climbs a hill, then looks down on a group of truly massive stones as in Figure 1.1. To the nineteenth-century poet and artist William Blake, Stonehenge was a symbol of raw power. The actual size of the biggest of the stones may be judged from Figure 1.2, but in Blake's imagination the stones expanded until they totally dominated the human stature. Figure 1.3 is a design from Blake's *Jerusalem*; in it the three human figures are utterly dwarfed by a great trilithon. It is of special interest for this book to note that in Blake's drawing the Moon has a bright crescent, as if it were in partial eclipse by the Earth. Remarkable too are the identities he intended for the three human figures: they are Bacon, Newton, and Locke, the masters of science and of reason. In an uncanny way, and in his own inimitable style, Blake seems to have guessed the modern astronomical associations of Stonehenge, indeed, the considerations which form the subject of this book. By its end, I hope to have convinced the reader that Blake was entirely right in choosing to show the Moon obscured, for it will be my purpose in the following pages to establish – I believe beyond reasonable doubt – that the purpose of Stonehenge was to *predict* the occurrence of eclipses.

The Druids had nothing to do with the construction of Stonehenge, which was completed a thousand years before the Celtic peoples invaded the British Isles. The work at Stonehenge was not done all at one time, as may be seen from the official ground plan of the whole of Stonehenge (frontispiece). The outer part

Figure 1.1. Stonehenge from the north-east. The white dots are concrete discs marking the positions of certain of the Aubrey holes (Chapter 2). The isolated standing stone next to the road along the bottom of the picture is the Heelstone (Chapter 2). (Courtesy of the Controller of Her Majesty's Stationery Office. Crown Copyright.)

Figure 1.2. The massive blocks of the central structure (Stonehenge III). The view is from Station position 94. This picture was taken before 1958, the trilithon 57–58 being at that time in a fallen position. The visitor's path was then in a pleasant grassy condition. (Courtesy of the Controller of Her Majesty's Stationery Office. Crown Copyright.)

Figure 1.3. Design 70 from the poem *Jerusalem* by William Blake.
(Courtesy of the British Museum.)

of the structure was probably begun before 2500 B.C. (see Chapter 2 for details), whereas the central structure shown in Figure 1.2 was built during a later period. Although the construction of this central structure was thought until recently to date from about 1700 B.C. to 1400 B.C., the new datings turn out

3

Figure 1.4. The still-complete portion of the Sarsen Circle viewed from the north-east. (Courtesy of the Controller of Her Majesty's Stationery Office. Crown Copyright.)

to be about five centuries earlier. In this chapter we shall be concerned with the central structure; the outer part will be the topic of Chapter 2.

The central structure is contained within a circle 97 feet in diameter known as the Sarsen Circle. A 'sarsen' is a particularly hard and durable form of sandstone, and the Sarsen Circle is so named because the 30 uprights, and the lintels which originally capped them, were made from boulders of this kind of stone. The Circle is still substantially complete throughout an arc of about 150° clockwise from approximately north. This still complete part of the Circle is shown in Figure 1.4.

Within the Sarsen Circle are three enormous standing trilithons, also made of sarsens. In the second millennium B.C. there were five interior sarsen trilithons arranged in a horseshoe formation, as in the reconstruction of Figure 1.5. Also within the Sarsen Circle was a subsidiary circle of smaller 'bluestones,' and within the trilithon horseshoe was a further bluestone horseshoe.

The landscape around Stonehenge is nowadays free from large boulders, as it probably was 5,000 years ago. Thus the

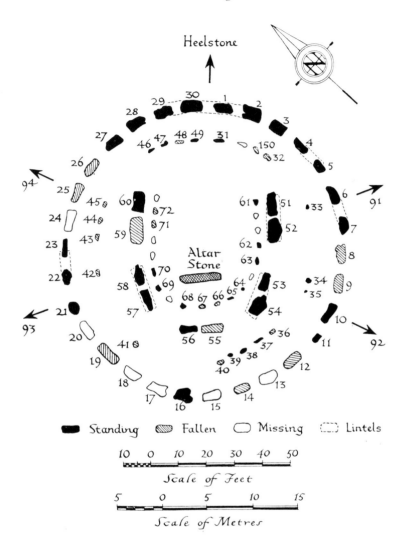

Figure 1.5. The original arrangement of stones in the structure of Stonehenge III. (Courtesy of the Controller of Her Majesty's Stationery Office. Crown Copyright.)

builders of Stonehenge could not simply put together local materials which happened to lie casually to hand. The stones had to be fetched deliberately, and in fact the bluestones were brought from a remarkably great distance, perhaps even before any construction work was done at all.* The bluestones taken as a whole are of varied mineral composition, with a peculiar mix of dolerite, rhyolite, volcanic ash, and sandstone. The late Dr. Herbert H. Thomas of the British Geological Survey established that this particular mix can be found only in a small area of south-western Wales; so these bluestones, weighing up to five tons each, must have been transported for 200 miles or more. Since it would be advantageous to use water-borne transport as far as possible, a likely route is that shown in Figure 1.6. The overland portions of this route, adding to about 20 miles, must have presented the main transportation problem.

In 1954, the British Broadcasting Corporation televised a programme in which concrete replicas of the bluestones were hauled overland by means of a technique judged to be within the capability of the men of the second millennium B.C., as shown in the artist's impression of Figure 1.7. The stones were each lashed to a wooden sledge, which was then placed on a bed of stout logs. The logs, serving as rollers, came free one by one as the sledge was hauled forward by many attached ropes. As a log thus came free at the rear, it was immediately carried to the front, where it again served as a roller. It was found that a team of some 20 strong lads could move a stone by this technique about one mile in a day, showing that the overland hauling of the bluestones was indeed possible, although hauling the 80-odd stones for 20 miles must have demanded a great effort, even from a large group of strongly motivated men.

The sandstone boulders which form the Sarsen Circle and the five trilithons present a still greater problem. It is thought that these stones came from near the town of Avebury, by the route shown in Figure 1.8. The most massive stones, weighing up to about 45 tons, were some 30 times heavier than the bluestone replicas used in the BBC experiment. Does this mean that 30 times as many men could shift the heaviest stones? Possibly,

---

*Suggested by R. J. C. Atkinson, *Antiquity*, XLVIII (1974), 62–3.

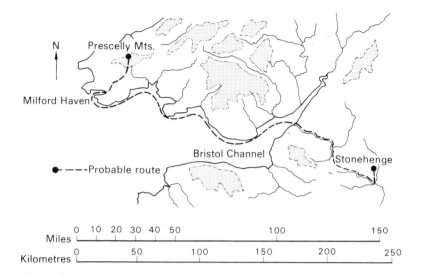

Figure 1.6. The likely route used for the transportation of the
bluestones from south-west Wales. (Adapted from R. J. C.
Atkinson, *Stonehenge*. Hamish Hamilton, 1956.)

but one can reasonably have doubts. Such a problem in
physics – and it is one – cannot automatically be solved by
merely multiplying the solution to a smaller problem, because
increased size and weight bring new factors into play. To avoid
increasing the strain in the ropes, 30 times as many ropes would
be needed; so men would constantly be getting in each other's
way. Men far from a sledge, pulling and tightening a rope, would
threaten to decapitate those nearer to the sledge. There would
be irritating problems of sideways motions. And worse, the
greatly increased weight would tend to increase all the frictional
forces, not just in proportion to the weight, but by an additional
factor caused by digging into the ground.

F riction is the essence of the problem. Imagine, for a moment,
that friction could be reduced as much as one pleased. Reduction to zero would not necessarily be the best arrangement.
Sledges to which heavy stones were lashed would then toboggan
on the merest downslopes, whereas on the uphill stretches of

Figure 1.7. An artist's impression of the neolithic method for hauling stones overland. (Courtesy of the Controller of Her Majesty's Stationery Office. Crown Copyright.)

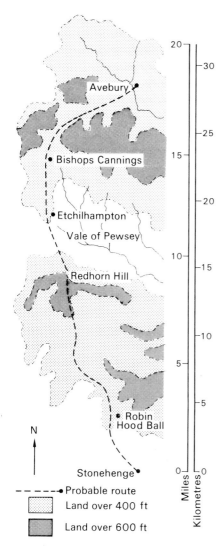

Figure 1.8. The likely route for the transportation of the Sarsen stones. (From R. J. C. Atkinson, *Stonehenge*. Hamish Hamilton, 1956.)

the route the men would never be able to stop and rest. A moderate degree of friction would therefore be desirable. I would guess the ideal situation would be one with an 'angle of friction' of about 2°. For most of the route there would be no tendency for the stones to run away, and on the steeper uphill slopes friction would not add seriously to the problem of hauling the sledges. On an uphill slope of 5°, for example, a pull of about 10,000 lbs would be needed to haul a 45-ton stone. If we reckon a sustained pull of 40 lbs a man, 250 men would be required – which is not a small number, even for what seems the best situation.

By the 'angle of friction', I mean the largest angle at which a sledge plus stone could rest without slipping backward. For this angle to be as small as 2°, friction would need to be only slight; less, I would think, than for the roller system used in the BBC programme. Professor P. Hill has suggested that one way to achieve such a low measure of friction would be to sledge the stones under hard-frozen winter conditions. The men themselves would then need suitable footwear, not for warmth alone, but also to obtain purchase on slippery ground. Skin-soled mocassins would probably be suitable.

Figure 1.9 shows roughly how average North American temperatures for the last 6,000 years have deviated from the current average. (I am not aware of comparable European data, but such may exist.) The long warm period from about 4300 B.C. to 2700 B.C. is known as the Climatic Optimum, when conditions in the British Isles were comparable to those now enjoyed in Mediterranean latitudes. Apart from one short warmer episode, the period from 2200 B.C. to about 1300 B.C. was distinctly colder than now, however. The suggestion that the sarsen boulders were transported in the depths of frozen winters around the year 2000 B.C. is thus consistent with the climatological data.

It is rare to find ancient monuments with stone cross pieces

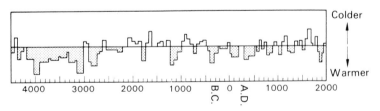

Figure 1.9. Climatic variations during the past 6,000 years. The estimated construction time of Stonehenge III, 2200 B.C. to 1900 B.C., lay in a cold period.

still in place – the standing columns now without cross pieces of ancient Greek and Roman temples come immediately to mind. Stone stands well in a vertical column, but only poorly in a horizontal span, because stone as a material is strong under compression but weak in tension. It is therefore remarkable to find some of the lintels still in place at Stonehenge, after as long as 4,000 years. The reason lies partly in the hardness of the stone itself, and partly in the way the Stonehenge lintels are supported. The upright stones, topped with the form of dome-shaped projection shown in Figure 1.10, were then fitted into holes made in the lintels, the method being like the mortise and tenon system used in joinery.

In addition to the shaping demanded by this mortise and tenon system, the boulders were trimmed into neat coffin-shaped blocks, especially the largest of the standing trilithons. The job was done by hammering stone on stone, another big task. Getting the stones into their final positions demanded the ingenious kind of operation shown in Fig. 1.11. The holes into which the uprights are fitted were dug with one wall sloped at an angle of about 45° and with the other three walls vertical. A stone would be gently eased down the sloping side of a hole, and then hauled and pushed into a vertical position, whereupon the hole would be quickly packed tight with soil and rubble. Getting the lintels into position was still more of a problem. In modern times, tall buildings are usually constructed from the bottom upward, but occasionally buildings have been erected from the top downward, the top being assembled first at ground level and then raised progressively as the construction proceeds. The lintels of Stonehenge may have been lifted in a somewhat similar fashion. Starting at ground level, they were probably 'pumped up' on wooden towers, which were built progressively as the stones were raised.

Turning back now to Figure 1.5, we can see that the horseshoe of 19 bluestones and the horseshoe of the five trilithons open up in the direction of midsummer sunrise. Marked out on a horizontal plane, and drawn for the latitude 51°11′ of Stonehenge, the direction of sunrise swings gradually from south-east

Figure 1.10. *Above:* The mortise and tenon joints can be seen on the tallest stone and its fallen lintel. *Below:* The tenon on the tallest stone is 9 inches high. (From R. J. C. Atkinson, *Stonehenge.* Hamish Hamilton, 1956.)

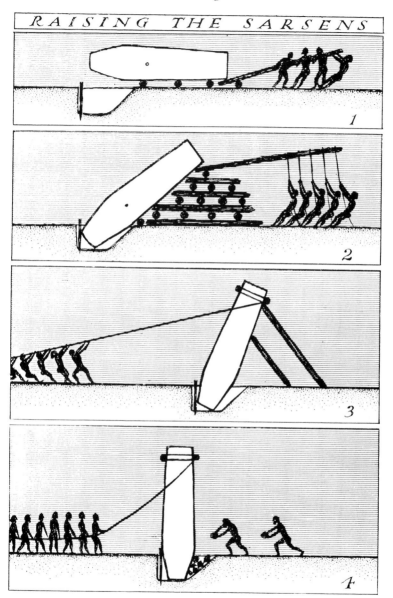

Figure 1.11. An artist's impression of the method used to erect the Sarsen uprights. (Courtesy of the Controller of Her Majesty's Stationery Office. Crown Copyright.)

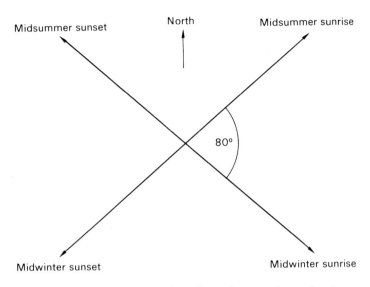

Figure 1.12. The annual swing of sunrise and sunset, drawn for the geographical latitude of Stonehenge.

in winter to north-east in summer. Then from summer to winter there is a return from north-east to south-east. This annual swing in the direction of sunrise (and of sunset) is shown in Figure 1.12. Simply by observing the direction of sunrise, one can follow the progression of the seasons of the year.

The close correspondence of the axis of Figure 1.5 with the direction of midsummer sunrise was noted by William Stukeley in the mid-eighteenth century. All commentators on Stonehenge since Stukeley have been aware of this remarkable fact, and most of them have taken it to imply a ritualistic connection with sun worship. A religious connection seems highly plausible, since it could well have provided a strong motivation for the construction of Stonehenge as a temple to the Sun.

Nobody, so far as I am aware, has argued that the axis of Stonehenge is only *accidentally* coincident with the direction of midsummer sunrise. Accidental coincidence appears to be ruled out by Figure 1.13, which is a sketch of the inner part of Sarmizegetusa, a prehistoric monument located near the village of

14

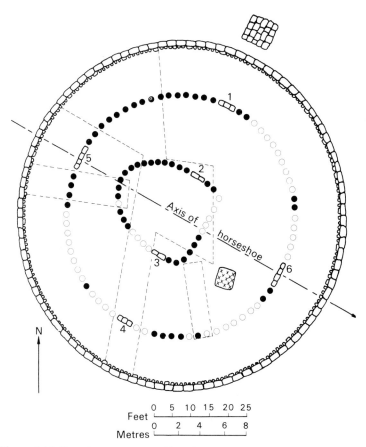

Figure 1.13. Sarmizegetusa, a prehistoric monument in Rumania. The axis of the central horseshoe of post holes points in the direction of midwinter sunrise. (From H. Daicoviciu, *Dacia* IV, 231.)

Grădiste in Rumania. The axis of the horseshoe of Figure 1.13 agrees with the direction of midwinter sunrise *as seen in Rumania*. Because of the difference in geographical latitude, the midwinter sun rises in different directions in Rumania and in England. Hence both Stonehenge and the horseshoe of Sarmizegetusa are appropriately oriented for their own particular sites – quite putting out of court, I would suppose, the suggestion that their associations with the Sun are due merely to chance.

How much of a connection with astronomy is implied by the relation of the axis of Stonehenge to midsummer sunrise? A minimum position would be to say that the primary motive for the construction of Stonehenge was ritualistic, and that a connection with the midsummer sun was made only for the subsidiary purpose of fixing the timing of a midsummer religious festival. A far more ambitious position has been taken, however, by Professor G. S. Hawkins in his book *Stonehenge Decoded*. According to Professor Hawkins, the archways of the five trilithons are positioned with respect to certain openings in the Sarsen Circle in such a way that they can be related to astronomically significant directions other than that of the midsummer sun, directions which involve the Moon as well as the Sun. On this second point of view, Stonehenge has a far closer connection with astronomy; indeed, so close that Stonehenge becomes as much an astronomical observatory as a religious temple.

In this first survey of the outstanding interior features of Stonehenge, it is scarcely possible to form an opinion on the merits and demerits of these two very different suggestions, although it is worth noting a difficulty with the first one. Returning to Figure 1.12, and remembering the 80° oscillation of the direction of sunrise between winter and summer, we must realize that there is no sudden jerk as the Sun turns around in its midsummer and midwinter positions. To the naked-eye observers of 2000 B.C., the Sun would appear to stand still for about a week in these positions. Thus a simple attempt to observe the midsummer sun would necessarily involve an uncertainty of several days. Such an ambiguity would scarcely be acceptable if one were seeking to fix a unique day for an important festival.

It is possible that the more subtle connections with astronomy noticed by Professor Hawkins had to do with overcoming this difficulty. To decide this question, we shall need to take account of certain features of Stonehenge which are less obvious than those which strike one on a first visit. The massive stones in the central regions, the ones originally laid out in the arrangement of Figure 1.5, are known as Stonehenge III. Predating this

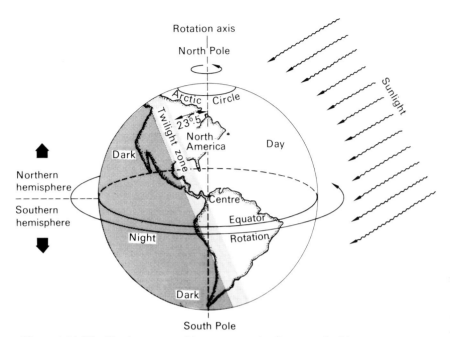

Figure 1.14. The Earth rotates with respect to the Sun once in 24 hours, which causes the sequence of night and day. Here it is dawn across the middle of the North American continent, on a day in the year close to 22 June (see Figure 1.15).

construction were Stonehenge I and Stonehenge II. The latter was a seemingly rather aimless affair, concerned with erecting the bluestones in two circles within what later became the Sarsen Circle. (The bluestones were subsequently rearranged by the builders of Stonehenge III.) Stonehenge I, on the other hand, appears to have been a deep and purposive construction whose significance we shall consider in subsequent chapters.

It will be useful, as an introduction to our consideration of Stonehenge I, if we end this chapter by discussing the reason why the direction of sunrise swings in the manner of Figure 1.12 between summer and winter. The spin of the Earth causes the cycle of day and night, as can be seen in Figure 1.14. The axis of the Earth's rotation is directed towards a point in the sky close to the North Star. This point can be considered to remain fixed during the course of the year – the fact that there is actually a very slow motion of the Earth's axis has no effect on any of the arguments in this book.

It is the Earth's rotation which causes the stars, the Sun, the Moon, and the planets all to rise in the east and to set in the west once each day. Because the rotation axis remains essentially fixed in the sky, we see each star rise and set always at the same places on the distant horizon. The situation for the Sun is different, however, because the Earth moves around the Sun during the course of the year. The direction of the Sun changes, therefore, with respect to the rotation axis, as shown in Figure 1.15. It is the changing direction of the line from the Sun to the Earth which causes the direction of sunrise to oscillate in the manner of Figure 1.12.

The position of the Sun can be measured by noticing the stars which lie close to the place on the horizon where the Sun rises. Making such observations from day to day, we find that the Sun's position moves relative to the stars. The reason for this change lies in the Earth's motion around the Sun, an effect illustrated by Figures 1.16 and 1.17. The constellations of Taurus and Gemini lie in the direction of the Sun at midsummer. After midsummer, the Sun passes in turn through Cancer, Leo, Virgo, Libra, and Scorpius, to reach Sagittarius by midwinter. Then from midwinter, the Sun continues through Capricornus, Aquarius, Pisces, Aries, and so back to Taurus by midsummer again. The particular constellations of stars which thus lie along the Sun's path in the heavens are known as the constellations of the Zodiac.

Few people nowadays will be familiar with these astronomical details, or indeed with the forms of other star constellations in the night sky. To the herdsmen and farmers of 5,000 years ago, however, such details were probably common knowledge. The groupings of the stars, the path followed by the Sun in the sky, and the path followed by the Moon were almost surely known with the same precision that a modern man reserves for the disposition of streets in his home town.

By 3000 B.C., Stone Age man had observed the sky for upward

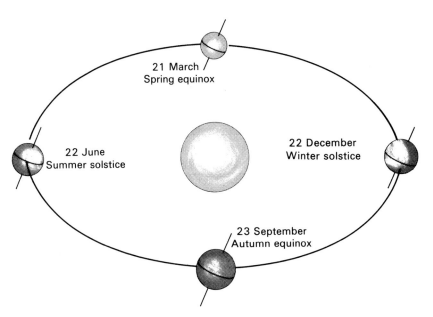

Figure 1.15. The Earth moves around the Sun in an orbit that is nearly a circle, and the axis of rotation of the Earth maintains an effectively fixed direction. The orientation of the Sun to the two hemispheres of the Earth changes during the year, and this change causes the seasons. The sizes of the Sun and Earth are both grossly exaggerated in this figure. (From J. C. Brandt and S. P. Maran, *New Horizons in Astronomy*, San Francisco: W. H. Freeman and Company. Copyright © 1972.)

of 10,000 years. It is reasonable to suppose that not only the astronomical details of the Sun's path, but certain more subtle variations of the Moon's path in the sky (still to be discussed), were discovered during this great interval of time, and hence were well known to the first builders when they arrived at Stonehenge.

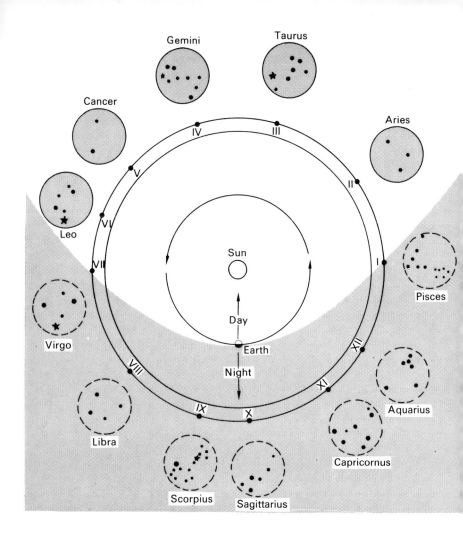

Figure 1.16. The positions of the twelve constellations of the Zodiac, marked along the ecliptic. The sizes of the Sun and Earth are grossly exaggerated. This is the position of the Earth at midsummer, and the constellations of Libra, Scorpius, Sagittarius, and Capricornus are visible in the night-time sky. The night-time sky is represented as a curve, to allow the constellations, which are actually very far away, to be brought close enough to the Sun and the Earth to be shown in the diagram.

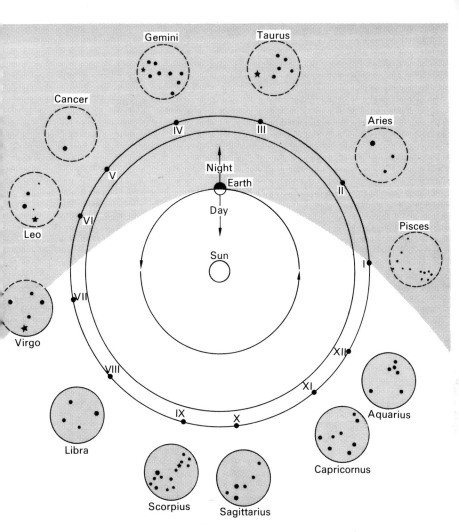

Figure 1.17. The position of the Earth at midwinter. The
constellations of Taurus, Gemini, and Cancer are visible in the
night-time sky.

## Bibliography

R. J. C. Atkinson, *Stonehenge*. Hamish Hamilton, 1956. (Also published as a Penguin paperback.)

R. J. C. Atkinson, *Stonehenge and Avebury*. Her Majesty's Stationery Office, 1959.

G. S. Hawkins, *Stonehenge Decoded*. Souvenir Press, 1966.

Ed. Sir Geoffrey Keynes, *The Complete Writings of William Blake*. Oxford University Press, 1969.

R. S. Newall, *Stonehenge: Department of the Environment Official Handbook*. Her Majesty's Stationery Office, 1959.

# 2

# The Sighting Lines
## of Stonehenge

*T*he first builders at Stonehenge began their work by digging a circular ditch, inside which they erected a bank some 20 feet wide and of about the height of a man. The bank has by now been eroded down to a height of only a foot or so, but it can still plainly be seen lying far out from the inner structure of Stonehenge III. The bank, with a diameter of 320 feet, much larger than the 97-foot diameter of the Sarsen Circle, was broken in the north-east, presumably to allow an uninterrupted view in the direction of the midsummer Sun. A distant stone,* the Heelstone, as it has been named in modern times, was erected in this direction.

The Heelstone, shown close up in Figure 2.1 and in relation to Stonehenge as a whole in Figure 1.1, is a sandstone boulder similar to the sarsens of Stonehenge III. It is about 20 feet long, about 8 feet wide, and about 7 feet thick towards its base. Its transport must have presented a problem like that posed by the large boulders of Stonehenge III. Unlike the builders of Stonehenge III, however, the first builders were not concerned with shaping the Heelstone by bashing and mauling it. The Heelstone is in its original form. At present, it leans at an angle of about 27° towards the centre of Stonehenge. If it stood upright originally, the top was some 2 feet higher than it is now, a detail that will assume significance in Chapter 4.

The chalk of Salisbury Plain is hard and not easily turned over. The technique used for loosening the surface was to

---

*A post, instead of a stone, may possibly have been used in the earliest times.

23

Figure 2.1. The Heelstone, showing the 27° tilt.

hammer the sharpened point of a deer antler into the ground, and then to pry loose the improvised pick – a hard job when thousands of cubic yards of material had to be shifted. Naturally, the bone antlers snapped from time to time, and were then thrown away as useless. Many such pieces of bone came to be buried by chalk debris at the site of the bank and ditch.

The Stone Age workman thus discarding a broken antler pick, doubtless with a curse muttered in a now-extinct language, could not have conceived in his wildest flight of fancy that the bony material would one day be dug up from its long resting place in the ground, and would then inform the men of the distant future of the times in which he himself was living. Yet it is so, and it happens like this.

The Sun and planets are embedded in a diffuse sea of highly energetic particles known as cosmic rays. Every second a few of these particles strike each square centimetre of the Earth's outer atmosphere. The central nuclei of atoms in the atmosphere are fragmented from time to time by collision with these incoming cosmic rays. Among the fragments are particles known as neutrons, which are quickly reabsorbed by the nuclei of ordinary nitrogen atoms, causing the nitrogen to disgorge a different particle, a proton, thereby changing the nitrogen to carbon,

$$\text{Nitrogen} + \text{Neutron} \longrightarrow \text{Carbon} + \text{Proton}.$$

The small quantity of carbon atoms so formed are different from the common kind of carbon, because the atoms, being derived from nitrogen, have nuclei each containing 14 particles instead of the usual 12. Physicists denote this difference by writing $^{12}C$ for common carbon and $^{14}C$ for the carbon arising from the effect of cosmic rays.

Chemical processes do not take much notice of the difference between $^{12}C$ and $^{14}C$. When plants absorb carbon dioxide from the air, it matters very little whether the carbon in the carbon dioxide is $^{12}C$ or $^{14}C$. Consequently a small fraction of the carbon present in plants is $^{14}C$. Because animals live by eating plants, or by eating other animals who do eat plants, the carbon in animals also contains a little $^{14}C$. The bones of animals thus contain $^{14}C$. In particular, the broken antler thrown away by our Stone Age workman contained $^{14}C$.

Now, whereas $^{12}C$ in a piece of wood or bone always stays

$^{12}C$, the $^{14}C$ changes slowly back to nitrogen. It does so in an interesting and precise way. After 5,700 years, one-half of the $^{14}C$ remains, the other half having changed back to nitrogen. After a further 5,700 years, that is, after a total of 11,400 years, one-half of the remainder has changed to nitrogen, leaving one-quarter of the atoms still as $^{14}C$. After a third period of 5,700 years, one-eighth of the original $^{14}C$ is left, and so on, in further characteristic time-steps of 5,700 years. This interval of 5,700 years is known as the 'half-life' of the $^{14}C$ atoms.

The ratio of the number of $^{14}C$ atoms to the number of $^{12}C$ atoms in a piece of charcoal, wood, or bone thus decreases as time goes on, and it does so according to a definite rule. If we know what the ratio was initially, and if we measure what the ratio is now, then the age of the bone or wood can readily be calculated. Such a procedure forms the basis of what is known as the $^{14}C$ (or 'radiocarbon') method for the dating of ancient objects and of ancient archaeological sites.

We understand now how our neolithic workman, when he threw away his broken antler pick, was unconsciously providing us with information about when Stonehenge I was built. Unfortunately, two complications stand in the way of a really precise calculation of the date when work began at Stonehenge. The sea of cosmic rays surrounding the Earth is constantly being disturbed by a wind of particles emitted by the Sun. Not only this, but changes in the Earth's own magnetic field can introduce fluctuations into the rate at which cosmic rays enter the terrestrial atmosphere. These disturbances vary on time-scales of centuries and millennia, which means that the production of $^{14}C$ by the cosmic rays has not been steady during past ages. As a result, we would underestimate the age of specimens of bone and wood that were formed at times when the production of $^{14}C$ was exceptionally high, and, conversely, overestimate the age of specimens formed when the $^{14}C$ production was exceptionally low. Unless suitable corrections are applied, errors of many centuries can arise in this way.

The age of wood in an actual tree can be found in another way, however: by counting the number of rings which the tree

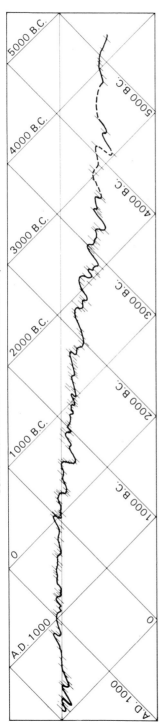

Figure 2.2. Bristlecone-pine calibration worked out by Hans E. Suess of the University of California at San Diego makes it possible to correct carbon-14 dates. The dates running across the top and the line on which they rest refer to carbon-14 dates in carbon-14 years; the dates running across the bottom and the lines on which they rest refer to bristlecone-pine dates in calendar years. The wavy line, which follows many individual measurements (the short slashes), shows how the carbon-14 dates go off with time. To calibrate a carbon-14 date, say, 2000 B.C., one follows the line for that date until it meets the wavy line. At that point a diagonal is drawn parallel to the bristlecone-pine lines, and the date is read off on the bristlecone-pine scale. The corrected date would be about 2500 B.C. (From Colin Renfrew, 'Carbon 14 and the Prehistory of Europe.' Copyright © 1971 by Scientific American, Inc. All rights reserved.)

has accumulated from year to year. Trees normally do not grow for more than a few hundred years, but Dr. C. W. Ferguson of the University of Arizona has discovered a bristlecone pine with the enormous age of about 8,000 years. This tree gave samples of wood whose age could be measured by the counting of rings, back as far as 6000 B.C. By measuring the ratio of $^{14}C$ to $^{12}C$ in these samples of known age, Professor Hans Suess of the University of California has been able to monitor the variable production of $^{14}C$ during past ages. With the variable production thus known, we can apply appropriate corrections to the conventional $^{14}C$ dating method. When we do so for the picks discarded by the workmen at Stonehenge, we arrive at an estimate in the range from 2500 B.C. to 3000 B.C. for the date of the ditch and bank of Stonehenge I. The method for correcting conventional $^{14}C$ ages is illustrated in Figure 2.2. The oscillations of the curve for corrected ages in the range 2500 B.C. to 3000 B.C. prevent the age determination of Stonehenge I from being pressed more closely.

There is also a second reason for not being too ambitious in seeking to give a precise date for the construction of Stonehenge I. It is assumed implicitly in the $^{14}C$ dating method that carbon has neither been added to nor taken away from the sample in question. Wood or charcoal stored in a dry climate meets this condition very well, but porous material like an antler pick, buried a foot or two below ground in the wet climate of Britain, does not. Contamination by organic material containing later carbon tends to increase the relative amount of $^{14}C$, thereby leading to an underestimate for the age of the sample. Since the antler picks found in the ditch of Stonehenge were probably thus contaminated to some degree, there is further uncertainty in the construction date, which is perhaps best left as somewhere within the period from 2500 B.C. to 3000 B.C.

It seems, then, that approximately 1,000 years elapsed between Stonehenge I and the last construction of Stonehenge III. The peoples responsible for Stonehenge I and Stonehenge III were almost surely different racially, with different languages, and probably with different customs and beliefs. This point needs emphasis because we ourselves tend to relate the two structures the wrong way around. Because Stonehenge III is more impressive physically, we tend to feel that the aims and motives of the builders of Stonehenge I were subservient to those of the builders of Stonehenge III. Actually, of course, the opposite is true. The builders of Stonehenge I were entirely unaffected by Stonehenge III, whereas the men who constructed Stonehenge III could well have been affected by legends and myths which had formed around distant memories of the builders of Stonehenge I. It follows that an interpretation of Stonehenge as a whole will be better sought in the earliest structure rather than in the later, more massive Stonehenge III.

So far, we have followed construction of Stonehenge I to the point where a ditch has been excavated, with the material dug from the ditch used to form a circular bank about 320 feet in diameter, the bank having a gap to the north-east to permit an uninterrupted view from the centre in a direction towards the Heelstone. Inside the bank, the workmen now proceed to dig a set of 56 holes at regular intervals around a concentric circle about 285 feet in diameter. The holes are named nowadays after the seventeenth-century antiquary John Aubrey, who mentions seeing regularly spaced depressions in the ground in his unpublished manuscript *Monumenta Britannica* (Bodleian Library, Oxford University). The Aubrey holes vary from strict positioning on the circle by no more than 19 inches, and the distances from one hole to the next vary by no more than 21 inches. It has been suggested that they may have been laid out by first dividing the circle into eight equal arcs, and by then subdividing each such arc into seven equal sectors. The second part of this operation would not be unduly difficult. Mark the beginning and the end of the arc to be subdivided. A rope with length equal to the radius of the circle is held tight between one assistant located at the centre and a second assistant at the starting point of the arc. The second assistant then walks until a shorter piece of cord is run out. A stone is placed at the position so reached, and the process is repeated until seven stones have been laid out. If the seventh stone falls short of the far end of the arc, the cord used in the operation is lengthened, but if the seventh stone goes beyond the end, the cord is shortened. The procedure is continued until the seventh stone falls at the far end of the arc.

The Aubrey holes average about 3 feet 6 inches in diameter, and about 2 feet 6 inches in depth. They were not dug to hold posts or upright stones. The filling material found in excavated holes varies greatly: chalky rubble, flints, wood ash, and crematorial remains. I would interpret this lack of uniformity of the filler material as an indication that the holes, or at least some of them, have been filled, or possibly refilled at times subsequent to their construction. (This point would affect the interpretation of $^{14}$C dates obtained from charcoal found in some of the holes.)

This digging of the 56 Aubrey holes raises interesting problems. Why no stones or posts? Why 56 holes? These are not the only puzzling questions which an observer of our own day and age might already have asked himself. At a first visit, Stonehenge III appears to be constructed on level enough ground, but Stonehenge III occupies only the inner part of the whole site. When examined on the scale of the bank and of the distant Heelstone, Stonehenge appears to be quite ill-chosen. The ground falls rather steeply towards the Heelstone, and the bank itself has an awkward tilt, being some four to five feet higher in the west than in the east. Given the wide area of Salisbury Plain, a more level site could have been found. Why did the builders accept these awkward level differences?

We shall arrive gradually at answers to all these questions. For the moment let us proceed to another crucially important feature, the four Station positions, marked 91, 92, 93, and 94 in Figure 2.3. Stones are found at two of these positions, the fallen one at 91 being about 12 feet long, and the top of that at 93 standing about 4 feet above ground. A filled hole, originally holding a stone or post, has been excavated at 92. Station 94 has not so far been excavated, however, and it has been marked on the official ground plan so that, together with 91, 92, and 93, a rather precise rectangle is formed, the short sides of which are close to being parallel to the direction from the centre to the Heelstone.

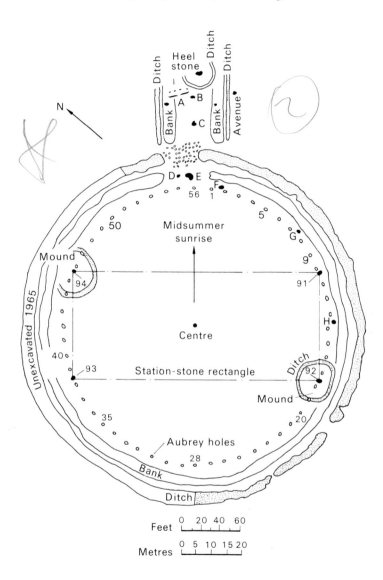

Figure 2.3. The structure believed to have been constructed by the first builders at Stonehenge. Positions 91, 92, 93, and 94 form the corners of a rectangle. The line from 91 to 94 very nearly bisects the line from hole 28 to the Heelstone. (Courtesy of the Controller of Her Majesty's Stationery Office. Crown Copyright.)

31

The larger stone at 91 is untooled, like the Heelstone, suggesting that the Station positions were established in the same period as the Heelstone. On the other hand, the stone at 93 has been shaped by hammering with smaller stones. This tooling has suggested to some commentators that the Station positions may have been added several centuries after the work of the first builders. A similar conclusion has been reached from the fact that certain small ditches that surround positions 92 and 94 have been cut through two of the Aubrey holes.* Both these arguments depend, however, on the debatable concept that what at first sight seems strange to us could not have happened. Why would anybody take the trouble to dig 56 holes, carefully spaced around a circle, and then put ditches through two of them? In Chapter 5 we shall arrive at a possible answer to this seemingly inscrutable question.

There is an interesting metrical equality which in my view establishes the Station positions as an important part of the Stonehenge I structure. To within the accuracy achievable by men working on the sloping terrain of the actual site, the line from 91 to 94 bisects the line taken from Aubrey hole 28 to a position at the front of the Heelstone, the 'front' being the side of the Heelstone facing towards the centre. This metrical equality suggests the manner in which the Station positions were laid out, namely, by running equal lengths of rope, one held at Aubrey hole 28, the other from a position at the front of the Heelstone, the ropes being increased in length until 91 and 94 on the Aubrey circle were reached. Position 92 was then taken diametrically opposite to 94, and position 93 was taken opposite to 91. This mode of construction obviously requires the short sides of the Station-position rectangle, namely 92 → 91, 93 → 94, to point in the direction of midsummer sunrise.

We next notice the important discovery of Mr. C. A. Newham, reported in the *Yorkshire Post* for March 16, 1963, that the long sides of the rectangle have a similar significance for the Moon. Like the Sun, the Moon pursues a path among the stars, a path

---

*Jacquetta Hawkes, *Antiquity*, XLI (1967), 178.

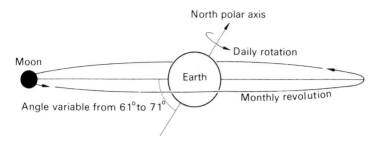

Figure 2.4. The sizes of the Moon and the Earth are grossly exaggerated in this figure, which has been drawn to show the relation of the Earth's polar axis to the plane of the Moon's orbit.

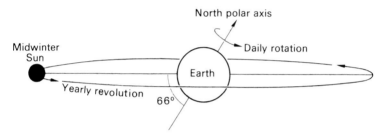

Figure 2.5. The relation of the direction of the Earth's polar axis to the plane of the path of the Sun in the sky. The size of the Sun and the size of the Earth even more are grossly exaggerated.

which is readily found simply by watching the Moon from night to night. This behaviour of the Moon is caused by the orbital motion shown in Figure 2.4. Just as the changing direction of the Sun in Figure 2.5 causes the yearly swing in the direction of sunrise, so the motion of Figure 2.4 produces a monthly swing in the direction of moonrise, as in Figure 2.6.

The situation for the Moon contains a complication, however. On a time-scale of years, or even of centuries, the yearly path of the Sun among the stars can be considered to be constant. But the path of the Moon changes slowly but discernibly from month to month, because there is a variation in the angle which the Earth's rotational axis makes with the plane of the Moon's

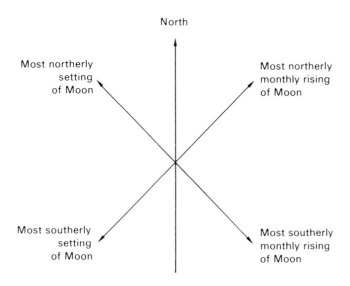

Figure 2.6. The monthly swing of the Moon is like the annual swing of the Sun in Figure 1.12.

orbit; it varies from 61° to 71°. This slow variation then causes a variation of the monthly swing of moonrise. When the tilt of the Earth's rotation axis is 61°, the monthly angle of swing of moonrise in Figure 2.6 is about 100°. When the tilt is 71°, the monthly swing is only 60°. These extreme possibilities are shown in Figure 2.7. If we observe the direction of moonrise, or moonset, during any particular lunar month of 27.32 days, we find a day-to-day swing of the form of Figure 2.6, with an angle somewhere in the range 60° to 100°. If we make such observations not just in one month, but for every month, we observe the angle of swing open gradually to 100°, then gradually shut down to 60°, then open up again to 100°, and so on. The time required for a complete oscillation of the angle of swing is 18.61 years, a period that will be found in later chapters to be of great significance.

The line drawn heavily in Figure 2.7 is almost exactly parallel to the long sides of the Station-position rectangle shown in Figure 2.3. This was the discovery made by Mr. C. A. Newham

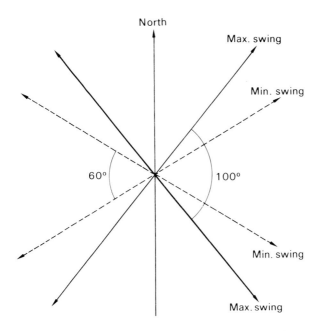

Figure 2.7. Because of the changing angle between the plane of the Moon's orbit and the Earth's axis of spin, the monthly swing of moonrise (and moonset) varies from about 60° to 100° at the geographical latitude of Stonehenge.

in his article appearing in the *Yorkshire Post*. Those who think the motive of the builders of Stonehenge was primarily ritual-istic try to pass off this relation of the Station positions to the Moon as being a fortuitous circumstance – which is surely implausible. Furthermore, Mr. Newham made a second re-markable point which is perhaps even more difficult to pass off as being due to chance. Suppose Stonehenge I had been con-structed at some other latitude. The same kind of ditch and bank, the same 56 holes around a circle, the same Heelstone could well have been used, although the direction from the centre to the Heelstone would need to be changed to match the direction of midsummer sunrise at the new latitude. None of this would make a problem for the builders. But what of the Station positions?

If the long sides, given by joining the Station positions, were to be related to the Moon in the same way, the Station positions would need to form not a rectangle, but a parallelogram with corners that were not right angles. Shifting Stonehenge only 50 miles to the north or south would change the required angles by as much as 2°. Obviously, if Stonehenge were moved very much in latitude, the rectangular construction of the Station positions would lose its relation to both the Sun and the Moon. In order for the Station positions to have such a relation at other latitudes, they would need to be set in a non-rectangular parallelogram that had precisely constructed corners, a task beyond the capacity of builders equipped only with sticks, ropes, and stones. The precise measurement of angles like 87° calls for a technique that would probably require the use of metal instruments.

Positions A, B, C, D, E, F, G, and H of Figure 2.3 have also been considered as belonging to Stonehenge I. Of these, A is an interesting row of four holes which held a set of wooden posts. Positions A, F, G, and H have been used by Professor Hawkins to obtain further alignments relating to the Sun and Moon, which are shown in Figure 2.8. These directions are the same as those shown in Figures 1.12 and 2.7. The numerical code of Figure 2.8 is related to the directions of Figures 1.12 and 2.7 by the equivalences set out in Table 2.1. The $\pm 24$ directions give the 80° swing of the Sun, the $\pm 29$ directions correspond

**Table 2.1**
*Interpretation of the numerical code of Figure 2.8*

| Numerical Code | Astronomical Interpretation |
|---|---|
| $+24$ | midsummer sunrise or sunset |
| $-24$ | midwinter sunrise or sunset |
| $+29$ | largest northerly monthly swing of Moon |
| $-29$ | largest southerly monthly swing of Moon |
| $+19$ | least northerly monthly swing of Moon |
| $-19$ | least southerly monthly swing of Moon |

to the 100° swing of the Moon, and the $\pm 19$ directions correspond to the 60° swing of the Moon.

## The Sighting Lines of Stonehenge

Figure 2.8 shows 16 astronomically related alignments, six of which involve the centre, and 11 of which involve one or more of the Station positions.* Professor Hawkins argues strongly that so many alignments cannot be due to chance, and hence

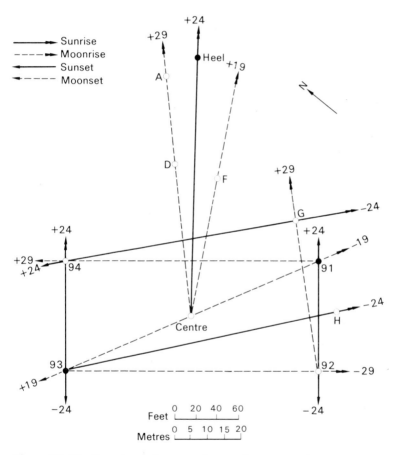

Figure 2.8. The lines shown here are close to the astronomically significant lines shown in Figures 1.12 and 2.7. (From G. S. Hawkins. *Nature*, October 23, 1963.)

---

*The reader will actually count 15 directed lines in Figure 2.8. Centre → A and Centre → D were given separately in Professor Hawkins' tables, making a total of 16 sighting lines.

37

that the relation of astronomy to Stonehenge is proven. This claim is equally strongly contested by others who have sought to weaken its statistical force, particularly by removing those lines which involve positions F, G, and H. Although holes certainly exist at these positions, the holes, it is argued, could have been formed by drainage channels or by the roots of trees, not being man-made at all. This view is unfortunately not directly open to test, because these positions suffered damage in a destructive excavation made some fifty years ago.

Criticism on these grounds appears invalid to me, since Professor Hawkins did not invent positions F, G, and H to suit himself. He used positions already mentioned in archaeological literature as having belonged (or having possibly belonged) to Stonehenge I, which surely he was entitled to do. If F, G, and H are really naturally made holes, their use is equivalent to throwing random numbers into the problem, a procedure which very rarely improves the apparent significance of the data.

I personally agree with Professor Hawkins that the alignments of Figure 2.8 are unlikely to be due to chance, but I do not obtain a probability of their being due to chance that is anywhere near as low as that which he finds in his book *Stonehenge Decoded*. Nor do I find that positions other than the centre, the Heelstone, and the four Station positions add much to the argument; so the controversy regarding F, G, and H appears to me to be statistically irrelevant.

My own analysis goes in the following way. Figures 1.12 and 2.7 contain between them 12 astronomical directions. Each astronomical direction defines an arc on the horizon in which a sighting line must lie if it is to be considered as 'agreeing' with the astronomical direction. The extent of the arc depends on how big an 'error' we are willing to permit between the sighting line and the astronomical direction. In Professor Hawkins' analysis, 'errors' up to about $\pm 2.5°$ are permitted; so each of the 12 astronomical directions covers a permitted angle of about $5°$ on the horizon. Taking all astronomical directions together, the permitted arcs add up to $60°$, which is 1/6 of the $360°$ of the whole horizon. Hence it follows that, for lines chosen at random, one in every six would appear to agree with one of the 12 astronomical directions. Now, joining the Heelstone, the centre, and the four Station positions two at a time gives 22 distinct directions.* On a random basis each of these directions would have a chance of 1 in 6 of agreeing with an astronomical direction. Now Professor Hawkins finds nine such agreements.† The chance of this happening in a random trial of the 22 directions works out at about 1 in 220, and the chance of there being 9 *or more* agreements works out at about 1 in 160.

My conclusion is that, although the chance that all the alignments are accidental is not exceedingly small, it is yet small enough for a *prima facie* case to be considered established. The case still needs backing by further investigation. Particularly, we need to consider the question of *why*. Why these astronomical alignments and not others? What were the builders seeking to *do*? If the astronomical interpretation of Stonehenge is correct, these questions must be answerable. Answers will indeed be suggested in the remaining chapters, answers that can be put to test, and so exposed to proof or disproof.

The measured alignments of Figure 2.8 have been criticized on the ground that Professor Hawkins took them from a plan of Stonehenge that was subject to distortion. The plan had a scale of 20 feet to the inch, and was supplied by Mr. B. V. Field of the (then) Ministry of Public Buildings and Works. One can think of many possible causes of error in the measured alignments:

(1) misjudgements of the archaeological positions themselves;
(2) inaccuracies in surveying the archaeological positions;
(3) inaccuracies of draftsmanship in making up the plan from the surveyed positions;

---

*Joining the six positions two at a time gives 30 pairings, but not 30 distinct directions. For example, centre → 91 gives the same direction as 93 → 91.

†I would also include 92 → 93 and 94 → 91, making 11 agreements and thus strengthening the argument.

(4) distortions in the reproduction of the plan;
(5) inaccuracy in measuring an assigned position in the repro-
    duction of the plan;
(6) inaccuracy in judging what point should represent certain
    of the positions – for example, the fallen stone at 91.

Of these, (2) and (5) should not have caused errors large enough
to be relevant. Concerning (5), Professor Hawkins did not use
a crude ruler-and-protractor technique for measuring
alignments; he used an instrument, of a type familiar to
astronomers, that is designed to measure assigned positions to
within a tolerance of less than 0.01 millimetre. As for (3), I would
suppose the Ministry of Public Buildings and Works would not
have issued the plan unless the draftsmanship was of first-class
quality. Certainly errors of the kind (1) are to be expected,
because recovery of the original positions after 4,000 to 5,000
years of wear and tear cannot be precise. Such errors could not
be helpful to Professor Hawkins, however, since any 'message'
present in the data would be garbled, not improved, by them.
Only because of (6) might errors 'favourably' directed have been
made. Professor Hawkins suggests that errors of this last sort
might have caused deviations of about $0.2°$ in the alignments of
Figure 2.8, well within the tolerance range of $\pm 2.5°$ discussed
above. I would also expect any distortions in the reproduction
of the plan to be much less than the tolerance range of $\pm 2.5°$.
Leaving aside errors of the basic kind (1), I find it surprising
that the directions measured by Professor Hawkins, and by the
assistants who helped him, should be questioned to within an
accuracy of, say, $\pm 0.5°$.

On one point alone, it seems to me, can the line of reasoning
used by Professor Hawkins be exposed to criticism: he tried to
fit the measured alignments not only to directions of the Sun
and Moon, but also to the risings and settings of the planets
and of the brightest stars. If one tries enough astronomical
objects (with suitably adjusted combinations), it is inevitable
that sooner or later some form of correspondence with the
measured alignments will be found – and this *would* result purely

from chance. It is for this reason that I referred to the case established by Professor Hawkins as a *prima facie* one, still requiring confirmation by other tests.

Because of my interest in Stonehenge, I am often asked to comment on other ancient monuments – Avebury, the Egyptian pyramids, Mayan temples, and so forth. It usually causes surprise when I express no interest. My indifference comes from a lack of definition of the problem. Given the plan of Avebury, it would be possible to invent many hypotheses about the purpose of this monument, so many indeed that sooner or later, just at random, the investigator would come on something which seemed to have a tolerable correspondence with the facts. For me, the case of Stonehenge was quite different. Thanks to the work of Mr. Newham and Professor Hawkins, the problem here is defined, particularly if we go on to add, as Professor Hawkins also did, the further assertion that the purpose of Stonehenge was to predict eclipses of the Sun and Moon. The problem is then defined with severe constraints. Without modifying the fixed structure in any way, can Stonehenge be used in a literal, practical way to predict eclipses successfully? Since the successful prediction of eclipses is a complex problem, an affirmative answer to this question would certainly not be expected unless eclipse prediction was indeed the purpose of Stonehenge.* The question is so definitive, permitting no room for manœuvre, that its investigation is quite explicit. The results of the investigation are described in the remaining chapters of this book.

Although the plan positions are subject to errors, probably of order $\pm 0.5°$ (for the reasons discussed above), I have not sought to change any of Professor Hawkins' measured values, since the testing of a hypothesis should not be confused by variations of the data. Nor in what follows will I attempt to extend the discussion to include more distant positions lying outside the area shown in the official ground plan. The possible significance of certain more distant positions has been discussed re-

---

*Anybody doubting this statement should try asking the same question for the Avebury monument.

cently by Mr. Newham, and by Professor Alexander Thom and his co-workers.

## Bibliography

C. A. Newham, *The Astronomical Significance of Stonehenge.* Leeds: John Blackburn, 1972.

Alexander Thom, *Megalithic Sites in Britain.* Oxford University Press, 1967.

Alexander Thom, Archibald Stevenson Thom, and Alexander Strang Thom, 'Stonehenge'. *Journal for the History of Astronomy,* V (1974), 71–90.

# 3

# *The Eclipse Interpretation*

*I*f the religion of the Stonehenge people was concerned with the worship of the Sun and the Moon as divinities, say, as a god and a goddess, a divine representation of man and woman, eclipses of the Sun and Moon would have been events of great importance. Successful predictions ahead of time would have conferred power and prestige on those who understood how the predictions might be made. The social conditions would therefore have been favourable to the development of astronomical ideas.

An attempt to explain the structure of Stonehenge in less complex terms, as a means of keeping track of the seasons, for example, seems much less plausible, since a much simpler structure would suffice for this simpler purpose. A small stone circle, of the kind found in many places in the British Isles, would be sufficient to provide a satisfactory seasonal calendar.

Eclipses occur when the Earth, Moon, and Sun are close to being in line. If the Earth lies between Sun and Moon, the Moon is eclipsed. If the Moon lies between Sun and Earth, the Sun is eclipsed, as shown in Figure 3.1. Hence, to be able to predict the occurrence of eclipses, we must know ahead of time what the directions of the Sun and Moon are going to be. For this, it is sufficient to consider the paths of the Sun and Moon as they appear to us on the sky, thinking of the sky as a distant sphere of very large radius. When we do this, the apparent paths intersect each other at the points $N$ and $N'$ of Figure 3.2. These points are often referred to as the *nodes* of the Moon's orbit. It is important to realize that the paths shown in Figure 3.2 do not represent the actual orbital motions. Figure 3.2 tells us

Figure 3.1. The solar corona at the total eclipse of June 1973.
(Courtesy of the High Altitude Observatory.)

nothing about the distances of the Sun and Moon – the Sun is, in fact, much more distant than the Moon. But this information we can afford to dispense with, since we are here concerned only with the directions of the Sun and the Moon.

The Sun's path in the sky can be taken as the same from year to year, but the Moon's path changes. The angle of $5°9'$ marked in Figure 3.2 stays the same, but the Moon's orbit slews around, causing the nodes $N$ and $N'$ to move along the Sun's path in the sense of the arrows shown in the figure. The nodes complete a circuit of the Sun's path every 18.61 years, which, it will be recalled, is the time required for the monthly angle of swing of moonrise to complete an oscillation between the extremes of $60°$ and $100°$ that were indicated in Figure 2.7. Indeed, it is just this slewing around of the Moon's orbit which causes the changing swing of moonrise and moonset. There is a connection between the two phenomena; so, by observing the changing

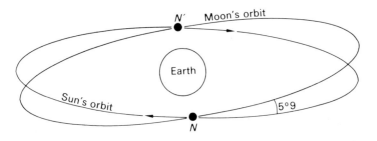

Figure 3.2. The paths of the Sun and Moon on the sky lie in planes which intersect at an angle of 5°9'. The points of intersection $N, N'$ of the paths are the nodes of the Moon's orbit.

swing of moonrise and moonset, one can deduce the whereabouts of the points $N$ and $N'$. These points are of critical importance in the eclipse problem, because an eclipse cannot occur unless the Moon is fairly near one or the other of them. At other times the Moon is well out of the plane formed by the Sun's path, and so the Earth, Moon, and Sun cannot then be in a straight line.

If this collinearity condition were strict – that is, if the centres of the Earth, Moon, and Sun had to be accurately in line – the Moon would indeed have to be precisely at one or the other of the two nodal points. However, because the Earth and Moon are bodies of appreciable size (as, of course, the Sun also is), eclipses can occur when there is a small departure from strict collinearity. In fact, the line from the Earth to the Moon can deviate from collinearity with the Earth–Sun line by up to about $1°$. Then, because the Moon's orbit is tilted at only a small angle of 5°9' to the Sun's orbit, the Moon can be up to about $10°$ away from a nodal point at the time of an eclipse. The situation is that a lunar eclipse occurs at full moon if the Moon and the Sun are each within about $10°$ of the opposite nodal points, and a solar eclipse occurs at new moon if the Moon and the Sun are within about $10°$ of the same nodal point.

In order to be able to decide whether or not eclipse conditions will occur at any time, say, during the next year, we therefore have to know: (1) where the Sun will be in its orbit at all times; (2) where $N$ and $N'$ will be with respect to the Sun's orbit; and (3) where the Moon will be. Let us discuss (1), (2), and (3) in sequence.

If we could measure the Sun's position at a particular moment, say, at midsummer or at midwinter, we could extrapolate the Sun's motion ahead, provided we know there are 365.25 days in the year. Since there are $360°$ around a circle, the Sun moves on the average by a little less than $1°$ per day. If we had a circle marked out accurately in degrees, we would simply move some pointer along the circle by rather less than $1°$ per day, starting the pointer at some agreed 'first point' on midsummer day. In fact, we would move the pointer by $360/365.25°$ per day. This would be rather an awkward fraction, but by graduating our circle with sufficient accuracy, we could manage it.

To obtain accuracy in graduating our circle, *we* would naturally make it in metal – but neolithic man could not do this. If we were forbidden to use metal, one recourse would be to make the diameter of the circle very large. But if we were to attempt to make a very large circle, say, from wood, how would we handle it? How would we prevent distortions in its shape? Since we have no wish to move the circle once it is suitably positioned, the obvious solution to the problem is simply to mark the circle out on a piece of flat ground.

Next we have to decide how to graduate the circle. We ourselves would be very likely to use the 360 units of division described above. But there is no reason why neolithic man would have used 360 units. Indeed, if Stone Age man were a sophisticated astronomer, there is every reason why he would *not* have used the number 360, because this number comes from bad astronomy. The interval between successive times of full moon is about 29.5 days. In a year of 365.25 days, there are about 365.25/29.5 such periods, which works out to rather more than 12. If we make the mistake of supposing that the year is exactly 12 such periods, and if we make the further mistake of supposing that each period is exactly 30 days, then the year comes out at 360 days, and there would be exactly $1°$ per day of motion of the Sun along our circle. But I will suppose that neolithic man had meticulously observed the Sun and Moon, not for just a few years, or even for a few centuries, but for many millennia. I will suppose that he knew perfectly well that there are 365.25 days in the year. Then he would *not* divide his circle into 360 equal parts.

It would, in any case, be cumbersome to attempt so many subdivisions of the circle. Fine graduation is only convenient when one works in metal. So let us choose a smaller number of division, say, 56. (The reasons for this number will appear shortly.) Now try the following rule. Move a marker, a stone on the ground, by two divisions every thirteen days. It will take $56 \times 1/2 \times 13$ days to move the stone around the whole circle, i.e., 364 days. At the end of a complete circuit, the time taken differs from a year by only 1.25 days; so motion along the circle according to this rule represents the motion of the Sun in its orbit to within an error of not much more than $1°$. This error is acceptable, because the rules which determine the occurrence of eclipses only require us to know the position of the Sun to well within $10°$. And provided we reset our stone with suitable accuracy every midsummer day, there will be no cumulative error piling up from year to year. Better still, if we reset our marker stone not just once a year but twice a year, on midwinter day as well as on midsummer day, the error will be halved.

Now we begin to understand why the builders of Stonehenge I dug holes at uniform intervals around a circle, and why they did not erect stones or posts in them. The holes were fixed reference points along the circle. The reason why there were 56 of these fixed points becomes still clearer when we consider the need also to keep track of the moving nodal points $N$ and $N'$. Two further stones, representing the positions of $N$ and $N'$, moved by three holes per year, would, as Professor Hawkins first pointed out, make a complete circuit in $56/3 = 18.67$ years, close to the 18.61 years actually required for the nodes to move around the path of the Sun. So the 56 Aubrey holes serve to follow both the Sun and the nodal points in a convenient way. Moreover, a marker moved by two holes each day completes a circuit in 28 days, not much different from the period of 27.3 days of the Moon in its path across the sky. Hence, by using the following rules, the directions of the Sun and the Moon, and the nodal points, can all be predicted ahead of time:

(1) Move a Sun marker, $S$, by one Aubrey hole each 6.5 days – that is, move $S$ alternately on the morning of the seventh day after an evening move, and on the evening of the sixth day after a morning move.

(2) Move markers for the nodal points $N$ and $N'$ by three Aubrey holes each year.

(3) Move a Moon marker, $M$, one Aubrey hole each morning and one Aubrey hole each evening.

The rule for the Moon is the least accurate of the three, and in Chapter 5 we shall see how by a suitable addendum to (3) it can be improved.

Since all the markers eventually go out of position, it is essential that they be reset from time to time. As we have already seen, $S$ could be reset at midsummer and at midwinter. The Moon marker $M$ could be reset at each full moon, when $M$ is opposite to $S$, and at each new moon, when $M$ is in the same direction as $S$. How then for the markers for $N$ and $N'$? If one understands the relation of the changing monthly swing of the Moon to the orbital picture of Figure 3.2, it is not difficult to prove that, in the month of the 18.61-year cycle when the Moon rises most to the north, the node $N$ must follow by 90° (and $N'$ must precede by 90°) the point where we place $S$ on midsummer day. Here 'follow' and 'precede' refer to the sense of the orbital motion of $S$.

But could the builders of Stonehenge I really have understood anything as sophisticated as this? Such a question cannot be answered by preconceptions. The answer must be sought by considering the evidence, as we shall do in the next chapter. For now, it is useful to notice that just the same rule applies to the most southerly rising of the Moon in the 18.61-year cycle. A second rule applies to the least northerly and to the least south-

**Table 3.1**

*Relation of the lunar nodes to the midsummer position of the Sun marker*
*(S) for the months of maximum ($\pm 29$) and minimum ($\pm 19$) swings of*
*moonrise*

| Numerical Code | Position of N | Position of N' |
|---|---|---|
| $+29$ | 90° following | 90° preceding |
| $-29$ | 90° following | 90° preceding |
| $+19$ | 90° preceding | 90° following |
| $-19$ | 90° preceding | 90° following |

erly swing of the Moon. We therefore have one rule for the outer swing of Figure 2.7, and another rule for the inner swing, with the relationships shown in Table 3.1. The numerical code used in Table 3.1 has been explained already in Table 2.1.

Four positions for $N$, $N'$, 90° apart from each other, are shown in Figure 3.3. The time required to go from one panel to the next is .25 × 18.61 years. The first panel gives the positions corresponding to the first pair of rows of Table 3.1, and the third panel gives the positions corresponding to the second pair of rows. Since the Sun must lie close to the direction of either $N$ or $N'$ at the moment of an eclipse, it is clear that when $N$, $N'$ have the configurations of the first or third panels, eclipses can occur only in spring and autumn. When $N$, $N'$ have the configurations of the second or fourth panels, eclipses can occur only at midsummer and midwinter. The times of the year when eclipses can occur thus follow $N$, $N'$ in their 18.61-year motion around the Sun's path, the path represented at Stonehenge by the Aubrey Circle.

On the day of a solar eclipse, sunrise and moonrise occur at essentially the same point on the horizon, as is clear from the requirement that the Sun and Moon shall be in the same direction on such days. On the day of a lunar eclipse, the Sun and Moon are in opposite directions. This condition requires sunrise and moonrise to be symmetrically distributed about the east–west line, as in Figure 3.4, because moonrise has the direction sunrise will have half a year later (or half a year earlier). For a lunar eclipse at midwinter, moonrise must thus be in the

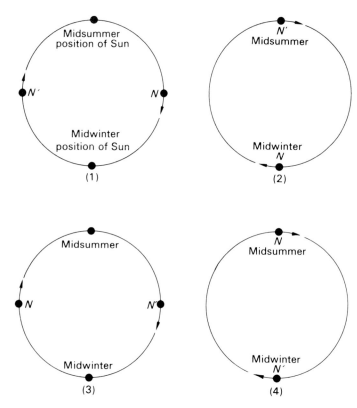

Figure 3.3. The nodes *N,N'* move around the Sun's path in 18.61 years. The four configurations of *N, N'* shown here occur ¼ × 18.61 years apart.

direction of midsummer sunrise – i.e., over the Heelstone. For a lunar eclipse at midwinter, moonrise occurs in the directions 93 → H, 94 → G. Eclipses never occur when moonrise is outside the 80° angle of the yearly swing of sunrise, the angle shown in Figure 1.12.

To recapitulate briefly: the eclipse interpretation of Stonehenge I has now explained the need for the sighting lines described in Chapter 2. It has also explained the need for the 56 Aubrey holes and why they were uniformly spaced around the Aubrey Circle. The eclipse interpretation will be further con-

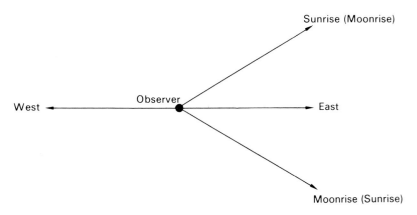

Figure 3.4. On a day of a lunar eclipse, the directions of sunrise and moonrise are symmetrically distributed relative to the east–west line.

sidered in the next chapter, where we will see that it can answer several other important questions about the structure of Stonehenge I.

## 4

# Tests of the Eclipse Interpretation

*T*he discussion in Chapter 3 required the position of the Sun
marker, *S*, to have been reset by observation of the Sun
twice each year, at midsummer and at midwinter. Yet the Sun
appears to 'stand still' on the horizon at these particular times
of the year, scarcely changing its rising and setting points. Any
direct attempt to find out which day is midsummer day by seeking to observe when the Sun rises most to the north would be
certain to fail, at any rate if attempted by naked eye with primitive equipment. A direct attempt to find midwinter day would
similarly fail. The likely error in the positioning of the Sun
marker would be as much as $5°$. With a similar error in the positioning of *N* and *N'*, and of the Moon marker, *M*, it would
become impossible to predict eclipses successfully. It follows
that, if the eclipse interpretation of Stonehenge is correct, the
sighting lines discovered by Newham and by Hawkins (the lines
of Figures 1.12 and 2.7) cannot have been used for measuring
the swings of the Sun and the Moon in a direct and obvious way.
How, then, could they be used?

To illustrate the problem graphically, suppose we attempt to
find the top of the symmetrical hill shown in Figure 4.1 by measuring its height directly. An error of measurement made near
the top inevitably makes the position of the top very uncertain,
as can be seen from the error band indicated in the figure. But
if, as in Figure 4.2, we look for two places of equal height on
opposite sides of the hill, and take the top to be midway between
them, similar errors of measurement are much less troublesome.

This idea can be adapted to the Stonehenge problem. Instead
of arranging sighting lines exactly on the directions of mid-

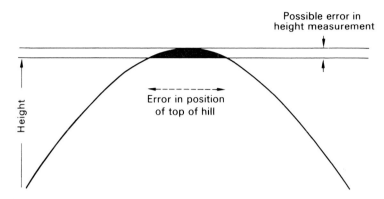

Figure 4.1. If we attempt to find the highest point of a symmetrically rounded hill by measuring its height directly, the inaccuracy arising from a small error can be quite large.

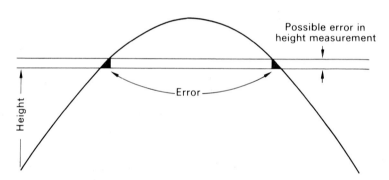

Figure 4.2. The effect of errors in measuring heights is much smaller when the top of the hill is taken to lie midway between two points of equal height on opposite sides of the hill.

summer and midwinter sunrise, arrange sighting lines that have a deliberate offset, as in Figure 4.3. The Sun will cross the stone that marks the more northerly dotted line of Figure 4.3 some days before midsummer, and it will recross the stone on its southward journey an equal number of days after midsummer. Midsummer itself is then midway between these two stone-crossing days. Midwinter is similarly found by using the southerly dotted line of Figure 4.3. In a similar way the times of the

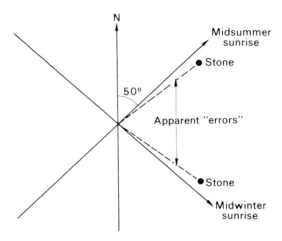

Figure 4.3. To employ the idea of Figure 4.2, the sighting lines to distant stones must lie *within* the angle of swing of sunrise.

largest and of the least monthly swings of the Moon, the angles of 100° and 60° in Figure 2.7, can be found.

Using this concept, we can subject the eclipse interpretation of Stonehenge to several stringent tests. Particularly, are there 'errors' of the type shown in Figure 4.3? When I first read Professor Hawkins' book *Stonehenge Decoded*, I was struck by the angular differences between the actual measured sighting lines and the astronomical alignments shown in Figures 1.12 and 2.7. Inadvertent constructional errors of 0.2° to 0.4° might have been expected, but not the errors of 1° to 2° that I actually found. With even primitive equipment, a stone could still be placed in position to within about a foot, giving an angular error, with the stone sighted from a distance of 200 feet, of no more than 0.3°.

The critical question arises, therefore, of whether these large 'errors' of 1° to 2° could have been made deliberately by the builders. To decide this question, note that random errors in the positioning of the two stones shown in Figure 4.3 would be just as likely to place each stone outside the angle of swing of sunrise as within it. The chance in a random situation that both stones will lie inside the angle of swing is one in four. There would be the same chance that both stones would lie *outside* the angle of swing, and a chance of one in two that one stone would lie inside and the other outside. Stones lying outside would obviously serve no astronomical purpose, since the direction of sunrise would never reach them.

With as many as eight different sighting lines associated with sunrise and sunset, we now have a severe test of the eclipse interpretation. Are the errors of the deliberate form required by this interpretation? Do the sighting lines always lie within the appropriate angles of swing? Or is the error distribution reasonably consistent with a random situation?

These questions can be answered from the data given in Table 4.1. The angles in the table are of the azimuthal type. Start by facing north. Then turn clockwise until the direction in question is reached. The angle thus turned is the azimuth. The calculated astronomical directions of the third column are those of the lines shown in Figure 1.12, and the measured directions of the actual sighting lines in Figure 2.8 are given in the fourth column. The above questions can be answered by comparing these calculated and measured values. But before we actually make the comparisons, we should note several factors that affect the numerical estimates.

The calculated directions involve the angle, $h$, of elevation of the actual horizon, which must be measured for the countryside around Stonehenge and for each direction separately. The calculated directions also involve the *meaning* of sunrise and sunset. Is sunrise the moment when the first tip of the Sun appears above the horizon, or when the centre of the Sun appears above the horizon, or when the whole disc of the Sun stands on the horizon? This question is answered below, and the answer to it has been incorporated into the calculated values of Table 4.1.

The measured values of Table 4.1 have been taken as directions to idealized points, judged to be the mathematical centres of the stones. Each stone has, of course, a certain width. The Heelstone, and the stones at 91 and 93, have rather flat tops, with widths of two feet or more. The width of a stone covers

**Table 4.1**
Relation of the calculated azimuthal angles of the solar alignments of Stonehenge 1 to their measured angles[a]

| Position | Seen from | Calculated (in degrees) | Measured (in degrees) | Required Condition |
|---|---|---|---|---|
| N.E. quadrant, elevation $h$ of horizon $\simeq 0.6°$ | | | | |
| 91 | 92 | $48.3 + 1.63h$ | 49.1 | Measured angle must be greater than calculated angle |
| Heel | Centre | $48.3 + 1.63h$ | 51.3 | |
| 94 | 93 | $48.3 + 1.63h$ | 51.5 | |
| S.E. quadrant, elevation $h$ of horizon $\simeq 0.7°$ | | | | |
| H | 93 | $129.2 + 1.63h$ | 128.2 | Calculated angle must be greater than measured angle |
| G | 94 | $129.2 + 1.63h$ | 129.4 | |
| S.W. quadrant, elevation $h$ of horizon $\simeq 0.6°$ | | | | |
| 92 | 91 | $230.8 - 1.63h$ | 229.1 | Measured angle must be greater than calculated angle |
| 93 | 94 | $230.8 - 1.63h$ | 231.5 | |
| N.W. quadrant, elevation $h$ of horizon $\simeq 0.3°$ | | | | |
| 94 | G | $311.7 - 1.63h$ | 309.4 | Calculated angle must be greater than measured angle |

[a]Because of the way sunrise and sunset are defined, there is a slight lack of symmetry in the calculated angles about the east–west line. Lunar parallax and atmospheric refraction have been included in the calculations.

the horizon for about $\pm 0.4°$ around the measured values set out in the table. Thus, even if the direction of the mathematical centre lies outside the angle of swing of sunrise or of sunset, part of the top of a stone could still extend inside the angle of swing.

The requirement for 'errors' to be considered deliberate – i.e., for the sighting lines to lie within the angle of swing of sunrise or of sunset – is given in the last column of Table 4.1. Inserting the appropriate $h$ values, we find that the required condition is satisfied in seven of the eight cases. Only for $91 \rightarrow 92$ does a discrepancy arise, and then only by $0.3°$ when stone width is allowed for, a small defect well within the *bona fide* errors involved in the measurements.

To assess the probability of such a situation arising at random, let us take the pessimistic position of regarding $91 \rightarrow 92$ as a failure. Then, with seven successes and one failure, the situation is like obtaining not more than one 'tail' in eight tosses of a coin. Some 20 trials, with each trial consisting of tossing a coin eight times, would be needed before there was an even chance of obtaining such a strongly biased result. Hence the odds that the 'errors' of the Stonehenge I sighting lines for the Sun are random are about twenty to one against. The distribution of the errors agrees very well, however, with the expectations of the eclipse interpretation.

It is important to realize that it was the logic of resetting the Sun marker $S$ which led to this understanding of the 'errors' of the sighting lines. We did *not* arrive at this result by trying a large number of hypotheses to explain the errors. Consequently the statistical argument has full weight. The disputed positions G and H appear in three of the sighting lines, always with 'errors' of the correct form, a circumstance which supports the view that these positions were deliberately man-made. The errors for holes made by trees would be distributed randomly, with only one chance in eight of obtaining the results of Table 4.1.

Let us continue by pressing still further the practical problem of making observations. What exactly do we mean by the Sun 'crossing a stone'? Simply gazing in a general way at the Sun and then deciding subjectively whether sunrise has occurred to the left or to the right of a line of the kind shown in Figure 4.3 will not do. An objective criterion for 'crossing the stone' is needed.

Suppose we arrange for our stone to project above the horizon. There will then be a small arc within which the stone obscures the first flash of the rising Sun. As midsummer approaches, the first flash of dawn would not be seen from the observation point for a few days as the Sun moves to the north, and after midsummer for the same number of days as the Sun swings back to the south. By observing both these 'blank' sequences of days, we can fix midsummer with clear-cut precision as lying half-way between corresponding days in the two sequences. How would we know that a particular day fell into a blank sequence? By having a colleague positioned elsewhere. He signals the first flash of sunrise. If we have not seen it from the appropriate observation point, the day is blank. A similar method can obviously be used for measuring midwinter, and also for the sighting lines for the Moon.

This interpretation of 'crossing a stone' permits no ambiguity in the definition of sunrise and sunset. Sunrise must be the moment when the tip of the Sun first appears. It cannot be the moment when the whole solar disc stands on the horizon, nor the moment when the solar centre appears on the horizon. Sunset is the moment when the Sun sinks wholly below the horizon. (This interpretation of sunrise and sunset was used in the calculations leading to the results of the third column of Table 4.1.)

Clear weather is needed to make such observations. Even during the Climatic Optimum, when the British Isles enjoyed a Mediterranean climate, there must still have been days of poor weather. If only a single sighting line were available for finding midsummer, one or another of the blank sequences might well be indeterminate in a particular year because of cloudy skies, and so the Sun marker could not be reset in that year. To reduce the chance of this happening, it would be sensible to construct several alternative sighting lines at different azimuthal angles, not only for midsummer but also for midwinter. The many sight-

Figure 4.4. A picture taken from the Centre, with eye position about 4 feet 8 inches above ground level. If the Heelstone were in a vertical position, it would stand about 2 feet higher, but ground level 5,000 years ago was also somewhat higher than it is today.

ing lines discovered by Newham and Hawkins show that the builders of Stonehenge I did precisely this. The multiplicity of sighting lines, so puzzling at first sight, has a natural explanation when we think through the process of observation in this way.

Indeed, thinking through the logic of observation has led us to a quite new prediction. Since stones that failed to project above the horizon would have been useless for establishing blank sequences, all sightings must have been to stones or posts that projected above the horizon. When seen from the centre, did the Heelstone project above the horizon? In modern times the top of the Heelstone breaks the north-easterly horizon as in Figure 4.4, for an observer about 5 feet tall standing at the centre. Raising the Heelstone into an erect position, which it presumably had originally, would give the same situation for an observer about 7 feet tall. Since the skeletal remains recovered from nearby burial sites show that neolithic man had a height close to 5 feet, the answer to our question would seem at first sight to be decisively affirmative. However, Salisbury Plain itself stood higher 5,000 years ago. Chalk, even hard chalk, dissolves slowly in water and is gradually washed away. Weathering of the land surface since 2500 B.C. must have lowered the eye level in relation to the top of the Heelstone. An affirmative answer to the question therefore requires that less than about 2 feet of surface chalk should have been dissolved by the rains of the past 5,000 years. Personally, I am optimistic that this was so!

The stone at position 93 is part of the solar alignment 94 → 93. When I walked to this stone, my heart sank. How could such a small stone, standing only 4 feet high, possibly project above the hill which lies to the west? The prediction that it should do so seemed disastrously wrong. With little hope, I set about investigating the matter. Since there is nothing on the ground to mark 94, I found out where this position was by referring to the official ground plan. First, I established a direction towards the fallen stone at 91, along a line passing about 3 feet outside the Sarsen Circle. Keeping to this line, I then paced off about 25 feet inwards from the central ridge of the bank. The position I reached turned out to be a hollow in the ground. From

Figure 4.5. The sighting line 94 → 93. Camera height about 5 feet 5 inches.

it I took the picture in Figure 4.5, which thus shows the present-day observing conditions for a person of my own height, 5 feet 9 inches.

It was noted already in Chapter 2 that the builders of Stonehenge I seemed little concerned to choose a level piece of ground. Now we see why quite marked slopes, like the slope across the Station positions, and like that towards the Heelstone, could be tolerated. The important relationships lay with the horizon, since the angle of elevation $h$ affects the astronomical directions, as can be seen from the third column of Table 4.1. What mattered was to choose a site with a smooth unbroken horizon which maintained an approximately constant value of $h$ in all directions. The relation of an observer to the height of a stone could always be adjusted, by choosing a stone of an appropriate size, or by scooping a hollow or raising a mound at the observer's position, or by adjusting the depth to which a stone was embedded in the ground. Indeed, the slope towards the Heelstone might have been considered an advantage, since it permitted a useful tolerance in choosing the length of the Heelstone. Had the stone been a foot or two shorter, or a foot or two longer, the same relation to the horizon could have been obtained by moving it either a little way up or a little way down the slope. Moderate variation in the length of the stone could thus be accommodated without the size of the Stonehenge I structure needing to be changed appreciably.

We saw above that the apparent 'errors' of the solar sighting lines of Stonehenge I are very likely to be deliberate offsets. The amounts of the offsets vary from about $1°$ to $2°$, except for $91 \rightleftharpoons 92$, where the offset, if any, can only have been of the order of the errors of measurement, say, $0.3°$. From a practical point of view, are these amounts well chosen? A sighting line $1°$ within the extremes of Figure 1.12 would be crossed by the Sun about 12 days before and 12 days after the moment when the Sun reached the extreme direction itself. Thus an observer using such a sighting line would complete his observations within a period of rather more than three weeks. For a sighting line $2°$ within the extremes of Figure 1.12, the period of observation would be some 40 per cent longer, or a little over a month, say, from early June to mid-July for the midsummer observations, and from early December until mid-January for the midwinter ones. Since it would be desirable to spread all the observations for all the sighting lines over several weeks, in order to avoid temporary spells of bad weather, the amounts of the offsets seem to have been excellently chosen.

**Table 4.2**
*Four lunar sighting lines that satisfy the required condition*[a]

| Position | Seen from | Calculated (in degrees) | Measured (in degrees) |
|---|---|---|---|
| G | 92 | 40.6 | 40.7 |
| A | Centre | 40.6 | 43.7 |
| 92 | 93 | 142.7 | 140.7 |
| 94 | 91 | 320.0 | 319.6 |

[a]These sighting lines are all concerned with the largest monthly swing of the Moon, and they have the same required conditions for each of the four quadrants as the solar alignments of Table 4.1.

When the lunar sighting lines of Figure 2.8 are analysed in the same way as the solar alignments, the result is not statistically impressive. Of the seven lunar sighting lines, the four set out in Table 4.2 meet the required condition, whereas the remaining three, given in Table 4.3, do not. The critic who wishes to can claim a weakening of the argument at this point. The eclipse interpretation is by no means destroyed by this situation, however, since the four lines in Table 4.2 can still be used for setting the nodal marker *N*. Even so, the eclipse interpretation demands that there be *no* discrepancies from the required condition, namely, that the sighting lines shall lie within the relevant angle of swing of the Moon or Sun. Hence the eclipse interpretation demands either that the three directions of Table 4.3 are false associations or that some new scientific consideration remains to be considered.

Let us examine the question of false associations. The probability argument of Chapter 2 showed that, in one case out of six, a purely arbitrary direction would happen by chance to agree with an astronomical alignment. Within the complex of the centre, Heelstone, and the four Station positions, there are 22 different directions. Among them two or three chance associations would therefore be expected. Accordingly, there is a sound statistical reason to expect that two or three chance

**Table 4.3**

*Three lunar sighting lines that do not satisfy the required condition[a]*

| Position | Seen from | Calculated (in degrees) | Measured (in degrees) |
|----------|-----------|-------------------------|------------------------|
| F | Centre | 60.1 | 61.5 |
| 91 | 93 | 122.3 | 117.4 |
| 93 | 91 | 300.3 | 297.4 |

[a]These alignments are all concerned with the least monthly swing of the Moon, and their required conditions are opposite to those in Table 4.1.

associations may be present in the list given by Professor Hawkins. Can we find non-statistical arguments to support the position that Table 4.3 contains such chance associations?

The differences between the measured and calculated angles are significantly larger for $91 \rightleftharpoons 93$ than for any other sighting line, being about $3°$ for $91 \rightarrow 93$ and $5°$ for $93 \rightarrow 91$. Indeed, these differences actually fall outside the tolerance range of $\pm 2.5°$ that was given in Chapter 2 as the criterion for forming the list itself. This is a rather clear indication that $91 \rightleftharpoons 93$ should not have been included.

There is, moreover, a strong geometrical reason for doubting that the $91 \rightleftharpoons 93$ diagonal of the Station rectangle was used as a sighting line. Looking back to Figure 2.3, we see that the Station rectangle is inscribed in the Aubrey circle in such a way that its shorter sides are almost parallel to the direction of midsummer sunrise. These properties do not define the rectangle uniquely. There are many ways in which such a rectangle might be constructed, and they allow a large range of values for the ratio of the length of the short sides to the length of the long sides. Among the many such rectangles, there will be one in which the $91 \rightleftharpoons 93$ diagonal does indeed coincide precisely with the $\pm 19$ lunar alignments (see the code definition of Table 2.1). The builders of Stonehenge might deliberately have used this criterion to construct the Station rectangle – but the evidence indicates that they did not.

We saw in Chapter 2 that the line taken from Aubrey hole 28 to a position immediately in front of the Heelstone is bisected by the line from 91 to 94. This fact strongly suggests that Station positions 91 and 94 were established by running out equal lengths of rope from the front of the Heelstone and from Aubrey hole 28. Only if the ratio of the radius of the Aubrey Circle to the distance of the Heelstone had been carefully adjusted by a sophisticated calculation would the resulting diagonal $91 \rightleftharpoons 93$ agree with the $\pm 19$ lunar alignments. In general, the Stonehenge sighting lines do not show evidence of this kind of premeditated calculation. The sighting lines usually have some freedom for later adjustment, which would be made simply by moving a stone (or post) a foot or two according to the outcome of trial observations. The only prearranged part of the geometrical structure appears to be the Station-position rectangle itself, which serves a dual lunar–solar role, because the choice of the latitude of Stonehenge itself seems to have been premeditated, as we saw in Chapter 2.

(At first sight, the fact that position G appears in both solar and lunar alignments in Tables 4.1 and 4.2 appears to be another example of premeditated calculation. But this dual property of G is really the same as that of the Station rectangle itself. Since G lies in the Aubrey Circle, with $92 \rightleftharpoons 94$ being a diameter of the circle, $92 \rightarrow G$ and $94 \rightarrow G$ must form a right angle at G.* Because the $\pm 24$ solar alignments and the $\pm 29$ lunar alignments are at right angles to each other at the latitude of Stonehenge, it follows that, if the position of G is chosen to give

---

*It is generally true, as a theorem in elementary geometry, that if two lines are drawn from the two ends of a diameter of a circle to any point on the circle, they must meet in a right angle to each other.

the correct solar alignment, the correct lunar alignment will be automatically guaranteed. This special property of G argues against the claim that G is of natural origin.)

It seems essentially certain, therefore, that the builders of Stonehenge did not plan to use the Station diagonal $91 \rightleftharpoons 93$ as a sighting line for the $\pm 19$ lunar alignments. It is accordingly reasonable to conclude that these directions should be omitted. We have then only two failures, namely, $91 \rightarrow 92$, Centre$\rightarrow$F among the remaining 13 solar and lunar alignments. On a random basis, this would be like obtaining only two 'tails' in 13 tosses of a coin, and the chance that this might happen is about 1 in 100.

But, as just stated, the eclipse interpretation permits *no* discrepancies. So what of $91 \rightarrow 92$, Centre $\rightarrow$ F? After we have allowed for stone width, the discrepancies in these cases are about $0.3°$ and $0.8°$, respectively. That for $91 \rightarrow 92$ lies within the uncertainties of measurement. That for Centre$\rightarrow$F would, I think, be rather outside these uncertainties. So a further consideration of Centre $\rightarrow$F is needed. The easy way to dispose of the problem would be to follow the archaeologists who regard F as a hole made by a tree root or by a drainage channel. Yet because we have found that there are good reasons for regarding G and H as man-made, I feel it would be somewhat artificial to adopt a different interpretation for F.

It is interesting that a change in the definition of moonrise could convert Centre $\rightarrow$F into a viable sighting line. In fact, there is another reason why the definition of moonrise must be reconsidered: I suspect that the first tip of the rising Moon would be difficult to observe in full daylight. One-half of all moonrises, and one-half of all moonsets, occur during daylight hours. Of course, the Moon is visible in daytime, provided we look for it deliberately, but it is then seen only as a pale object, not dominating our attention as it does when it appears bright in the night sky. Recognizing the first tip of the rising Moon, which our blank-sequence technique requires us to do, would be difficult by day. Supposing one knew exactly where to look, and at what time, then I think a person blessed with good sight might spot the rising Moon after the upper limb had risen a few minutes of arc above the horizon. The place of moonrise would indeed be known to the Stonehenge observers, but without a clock to judge the exact time of day, they would have had to maintain a careful watch for an uncertain period – for at least half an hour ahead of moonrise, I would say. Such a long anticipatory wait, staring keenly at the horizon the whole while, would unquestionably have been exceedingly fatiguing. The same difficulty does not occur at night, of course, because the rising Moon then makes an easily visible glow in the sky, which brightens noticeably just before the first tip of the lunar disc appears on the horizon.

To see how failure to observe moonrise during daytime could cause some difficulty, consider how we propose in theory to establish a blank sequence for the most northerly monthly swing of the Moon. (Blank sequences for the most southerly monthly swing, and for the least northerly and the least southerly swings, have a similar difficulty.) As the monthly swing of the Moon opens up from 60° to 100° (see Figure 2.7), moonrise eventually occurs in the blank arc established by a stone or post, on at least one day in the month. If we can establish by our nightly observations when this first happens, we know that the month in question belongs to the blank sequence for which we are looking. As the angle of swing continues to open out, moonrise eventually becomes visible to the north of the stone or post, on at least one day in the month. If we can discover this other day, we know that the blank sequence of months has ended. But if we fail in practice to observe moonrise on a critical day, or days, for the reason that the Moon is in the daytime sky, we will be uncertain whether or not a particular month belongs to the blank sequence. The outcome of this uncertainty would be an error of one, or more, months in the positioning of $N$ and $N'$; by about 1.6° for an error of one month, by 3.2° for an error of two months. Although such errors are just tolerable, they would be irritating enough to prompt a search for a new method in which daytime risings of the Moon were made readily observable and subject to measurement.

Suppose the Moon to have risen until the whole disc, not just the tip, stands on the horizon. What happens next is that the Moon moves up into the sky along a path which makes a rather shallow angle with the horizon, say, 25° to 30°. Consequently a stone or post projecting above the smooth horizon could obstruct a part of the disc of the rising Moon as in Figure 4.6. The observer of this phenomenon would not be required to wait an agonizingly long time, gazing steadfastly at the horizon waiting for the first tip of the Moon to appear. The observation would begin only after the Moon was clearly and easily seen to be rising.

A critical situation, suitable for measuring a particular monthly swing of the Moon, is needed, since nothing precise could be deduced from the general situation shown in Figure 4.6. Such a critical situation could arise, however, when the Moon is just grazing the stone or post, as in Figure 4.7. Notice that a similar idea could never be used for the Sun. The Sun is too overwhelmingly bright for a naked-eye observer to look successfully for the grazing-contact situation of Figure 4.7.

Interpreted in this way, the calculated angle for Centre →F is $61.1° + \sim s$, where $s$ is the elevation of the stone or post above the horizon (in degrees). The approximation symbol ($\sim$) appears here because we can estimate $s$ only roughly; a precise estimate would depend on our knowing the exact shape of the obscuring stone or post. It is reasonable to take $s$ to lie in the range 0.3° to 0.5°; then the calculated angle for Centre →F exceeds the measured angle, and this sighting line becomes a viable +19 lunar alignment. Indeed, the value of 61.1°, changed from 60.2° by the changed definition of moonrise, is so close to the measured 61.5° that it lies within the errors of measurement.

Since the grazing-contact method of Figure 4.7 could be used both at night and during the day, it might seem as if all the lunar alignments would best be used in this way. However, so long as the critical occasions of moonrise occur at night, the blank-sequence method is the more accurate method, and high accuracy would be of main importance. For the nodal marker $N$

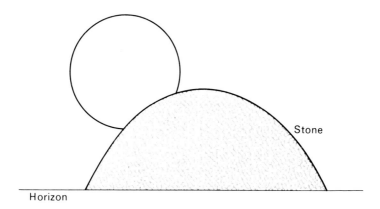

Figure 4.6. The rising Moon is partially obscured by a stone projecting above the horizon.

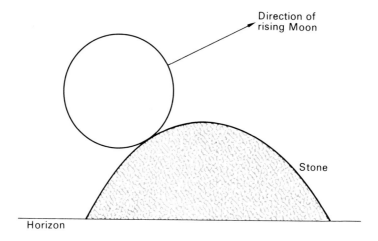

Figure 4.7. Grazing contact of the rising Moon with a distant stone.

to be set within a tolerance of $1°$, the azimuthal position of the rising Moon must be defined to within less than $0.1°$.* This can be done with night-time risings of the Moon by the blank-

---

*The reader interested in details will find an analysis of errors and tolerances given in section 7 of the appendix.

sequence method, but I doubt that the method of Figure 4.7 could achieve this measure of accuracy. The method of Figure 4.7 would be a fall-back position, to be used only in the event of unluckily placed daytime risings of the Moon.

For the Sun, depending on how much the sighting line is off-set, and on the size of the stone used, a blank sequence might last for three or four days, or even for a whole week. For the Moon, the blank sequence during which the monthly extreme of moonrise (or moonset) was not seen would last very much longer. Indeed, for a stone with an offset of $2°$, producing a blank arc of $1°$, the monthly extreme of the Moon might not be seen for a whole year. This situation would surely be highly in-convenient, since an observer, in order to establish the end of a blank sequence, would need to keep watch month by month for a very long time. It would, I think, be most convenient if a blank sequence were to last for two or three months only. For this to happen, the blank arc must be small, much less than $1°$, much less than the blank arcs permitted by the solar alignments. This requirement would demand that either posts or stones with narrow tops be used in the lunar alignments – no more than about 10 inches for a sighting length of 200 feet. Of the five lunar sighting lines (92→G, Centre →A, 93→92, 91→94, Centre →F), the hole at A has the cylindrical shape characteristic of a post-hole. The holes at 92, F, G, have irregular cross-sections and so probably held stones. Unfortunately no stone is now found at 92, F, G, or 94, and so nothing can be said about the widths of the tops as they were originally.

Why not also use posts for all the alignments? I see no scien-tific reason why not; indeed, the first builders at Stonehenge might well have used posts rather than stones. It is true that stones are less likely than wooden posts to be accidentally dis-placed, but this advantage hardly seems great enough to offset the difficulty of hauling large boulders to the site. The use of stones where possible may have been dictated by religion and ritual. Stones have a powerful emotional effect – I doubt that Stonehenge would receive its present-day throng of visitors if the whole structure were in wood instead of stone.

The builders of Stonehenge I seem to have accepted a free-standing position for the observer, since no attempt was apparently made to place a constraint on the observer's eye position. I suppose if one were required to stand every day at a place marked by a stone or post, it would be possible to maintain a constant eye position to within a tolerance of two or three inches. In sighting a stone at a distance of 150 feet, there would be inadvertent day-to-day variations of line by rather less than 0.1°. Now, two weeks or so before midsummer, the tip of the rising Sun moves along the distant horizon at an angular rate of about 0.1° per day. So personal errors of eye position would not usually cause trouble in observing blank arc days. Occasionally, however, a day would be lost or gained because of such errors, leading to a mistiming of midsummer day by half a day (an error of a day would be divided by two, because of the before-and-after method of identifying midsummer day). The resulting error in the placing of the Sun marker on the Aubrey Circle would be about 0.5°. Such an error is acceptable within the eclipse interpretation, since it is comparable to that which arises inevitably from the slight inaccuracy of the rule for moving the Sun marker $S$ (Chapter 3).

These considerations show why Stonehenge I could not have been made much smaller, Distances of the order of 150 feet are necessary in a free-standing system in order to prevent personal errors from introducing serious inaccuracies into the settings of $S$ and $N$. Stonehenge I could have been made larger, but more labour would then have been expended in its construction. The scale used was chosen to give satisfactory accuracy with the least effort.

Now, the scale of Stonehenge III is distinctly smaller than that of Stonehenge I, the radius of the Sarsen Circle being somewhat less than 50 feet, about one-third of the radius of the Aubrey Circle. Hence the use of Stonehenge III for eclipse prediction would require the observer's eye position to be constrained so that day-to-day variations were reduced from about 3 inches to 1 inch or less. How was this done?

The trilithon numbered 51–52 in Figure 1.5 was constructed

with a narrow V-shaped opening between the uprights. The observer can fit himself quite tightly into this opening, thereby fixing the eye position closely. We should notice here that the observer is never required, for any of the Stonehenge sighting lines, to take up an absolutely specified position. What matters is that a position, once chosen, should be maintained the same from day to day. An observer using the V-shaped opening of 51–52 could therefore choose to wedge himself in his own way. I would suppose an observer of 4,000 years ago could likewise wedge himself in the openings of the other four trilithons shown in Figure 1.5, although this supposition cannot be verified directly, since stones 55 and 59 are not at present in standing positions, while 53–54 and 57–58 have been seriously affected by weathering.

By relating the gaps between the uprights of the Sarsen Circle to the openings of the trilithons, Professor Hawkins has compiled a list of possible astronomical alignments. These are shown in Figure 4.8. With the tip of the rising Moon or Sun taken to define moonrise or sunrise, these suggested sighting lines have been analysed in Table 4.4, as was done in Tables 4.1 and 4.2 for the sighting lines of Stonehenge I. The horizon elevations given in Table 4.1 have been used to obtain the calculated angles.

Table 4.4 does not include the direction in Figure 4.8 towards the Heelstone, since this direction is not exclusive to Stonehenge III. Nor does it include any 'required condition', because no such test is possible here, there being no stones (except the Heelstone) in the centres of the Sarsen gaps.

It will be seen from Table 4.4 that the tolerance between measured and calculated values has increased considerably from the $\pm 2.5°$ used in Chapter 2 for the sighting lines of Stonehenge I. Indeed, for one direction the measured and calculated values differ by more than $8°$. With this widening of the tolerance range, the chance that an arbitrarily chosen direction will happen to 'agree' with one of the 12 astronomically significant alignments is much larger than the value 1/6 used in Chapter 2 for the sighting lines of Stonehenge I. Yet even for a chance as high as 1/2, the associations of Table 4.4 are still significant,

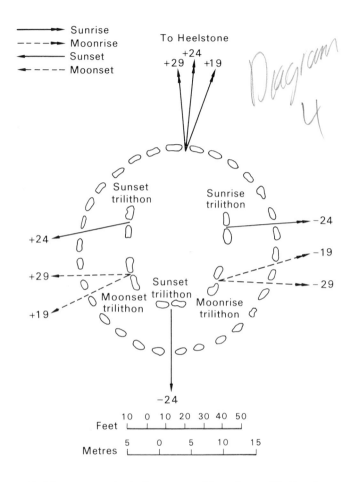

Figure 4.8. The astronomical alignments of Stonehenge III. For the numbering of the stones, see Figure 1.5. (Courtesy of the Controller of Her Majesty's Stationery Office. Crown Copyright.)

even though there is also some question whether two of the alignments should properly be included. If the openings of all the trilithons were originally as narrow as that of 51–52, a tightly wedged observer in 53–54 would have had difficulty in seeing through the Sarsen gap 8–9. Similarly, there would be difficulty in using 57–58 for more than one Sarsen gap. If the gaps in all

**Table 4.4**
*The suggested astronomical alignments of Stonehenge III*

| Position | Seen from | Calculated | Measured | Sighting Line for |
|---|---|---|---|---|
| 8–9 | 53–54 | 122.3° | 120.6° | Moonrise |
| 6–7 | 51–52 | 130.3° | 131.6° | Sunrise |
| 9–10 | 53–54 | 142.8° | 139.4° | Moonrise |
| 15ª–16 | 55–56 | 229.8° | 231.4° | Sunset |
| 20ª–21 | 57–58 | 300.3° | 292.0° | Moonset |
| 23–24ª | 59–60 | 311.2° | 304.7° | Sunset |
| 21–22 | 57–58 | 320.0° | 315.2° | Moonset |

ª These are missing stones whose positions have been estimated from neighbouring stones.

five trilithons were originally as narrow as 51–52, I would say that only a single opening in the Sarsen Circle would have been conveniently oriented for each trilithon, making only five directions available for observation. The fact that each of these five directions happens to 'agree' with an astronomical alignment seems to indicate that the trilithon positions were deliberately chosen in terms of the alignments. For a single trilithon with a narrow V-shaped opening placed at random within the Sarsen Circle, there would be a chance of about 1 to 2 that the line through the associated Sarsen gap would happen to 'agree' with one or another of the 12 astronomical directions. For a single trilithon such an 'agreement' would be quite unimpressive. But the 'agreements' of Table 4.4 for all five trilithons *is* impressive – the chance that this situation could happen at random is only 1 in 32. Hence it seems likely that the builders of Stonehenge III placed the trilithons in relation to the Sarsen Circle according to their astronomical associations.

But of what use were these associations? An observer standing in the V-shaped opening of a trilithon, wedged to look through the corresponding gap of the Sarsen Circle, sees as much as 10° of the distant horizon. Nothing of quantitative scientific value could be achieved from observations made through the centre of such a gap, since there is no adequate reference point there for judging day-by-day variations of sunrise or of moonrise. Only if the Sarsen uprights themselves were used to establish the critical appearance of the first tip of the rising Sun or Moon could an observation of scientific value be obtained. For example, considering the solar alignment 51–52→6–7, there will be a day, about three weeks before midwinter, when the tip of the rising sun is first seen appearing from behind Sarsen upright 6, and after midwinter there will be a reversed sequence, with the tip of the rising Sun moving northward across the gap, until it disappears again behind Sarsen upright 6. By noting both the appearance and the disappearance of the Sun in this way, one could calculate the moment of midwinter precisely as the day halfway between. The other alignments could be used similarly.

It is interesting that the lunar alignment 57–58→22 (i.e., to Sarsen upright 22, not to the gap between uprights 21 and 22) would have a measured angle close to 320°, close not only to the calculated direction but also to the sighting line 91→94 of Stonehenge I. I think it unlikely that these near equalities are fortuitous.

Let us summarize the work of this chapter. Quite apart from the suggestive circumstantial aspects of the astronomical alignments discovered by Newham and by Hawkins, we have come on several tests that are essentially predictive. In particular, we have seen how great are the odds against the possibility that the apparent 'errors' in the positioning of the stones are random in origin. No documentary evidence about the distant past could be so strong. Documents are frequently wrong, sometimes because of inadvertent errors, sometimes by design. For Stonehenge, on the other hand, it is implausible to argue that a people ignorant of astronomy chose positions for the stones that happened by chance to display great astronomical subtlety! It might be hard for a historian to accept the idea that a geometrical arrangement of stones and holes can provide evidence much stronger than that of a document, but I believe this to be true.

Although misgivings have sometimes been expressed concerning the accuracy of the archaeological positions themselves, the present argument strongly suggests that these positions are

singularly accurate. That they should be so, after all the generations of people who have pulled and pried and dug and rooted,
is admittedly remarkable. One might have expected that stones
moved, stones taken from the site, stones collapsed by the
ravages of wind and water, would all have conspired to garble
the message. The wonder of it is that the message is still there,
almost as clear as it was in the beginning.

# 5

# Making Predictions

*In* Chapter 3 we saw that, in theory, eclipses can be predicted ahead of time by following certain rules for moving markers representing the Sun, Moon, and the nodes of the Moon's orbit. A practical attempt to apply these rules in detail would inevitably raise the following difficulties.

(1) Although the procedure for updating the markers in real time would be straightforward in its execution, an attempt to predict ahead of time would be a complex affair, requiring many instructions to be transmitted across the Aubrey Circle. Even with a carefully worked out drill for moving the markers, it would be difficult to avoid misunderstandings. Shouting instructions across a distance of about 100 yards would lead almost inevitably to confusion.

(2) According to the rules of Chapter 3, the markers $S$, $M$, $N$, $N'$ are moved in discrete steps of one Aubrey hole at a time, whereas the actual Sun, Moon, and nodes all move smoothly with respect to time. At each jump the markers correct for a delay of one Aubrey hole, so that, immediately before a jump is made, a marker is out of true position by the angular separation between adjacent Aubrey holes, i.e., by $360/56 = 6.43°$. This lagging and jumping would be of little consequence when the markers were remote from an eclipse configuration, but close to an eclipse configuration it would be essential to refine the procedure for following the Sun, the Moon, and the nodes. It would be necessary to subdivide the $6.43°$ steps from one Aubrey hole to the next; so there would need to be still greater complexity in the drill for moving $S$, $M$, $N$, and $N'$.

(3) At the end of a predicting session, a precise return to the

starting configuration would be crucial, since the real-time settings of the markers would otherwise be lost.

Whereas (1) and (2) are merely difficulties to be overcome, problem (3) presents a serious danger, since the settings of $N$ and $N'$, if ever lost, could take several years to recover. Unless a record of the current setting were kept available, a mistake would inevitably occur sooner or later. A record could be kept by having two sets of markers, two heavy stones to keep the real-time positions of $N$ and $N'$, and two lighter stones to be moved in attempts to predict eclipse configurations ahead of time. Likewise, two stones, one heavy and one light, could be used for $S$ and also for $M$. Although this duplication of the markers overcomes difficulty (3), it prompts the question, why attempt to make the predictions at Stonehenge itself? Why not make up a pegboard of the form of Figure 5.1, keeping Stonehenge always as a real-time record of the positions of $S$, $M$, $N$, and $N'$?

In Chapter 3 it was considered that the accurate marking of as many as 360 divisions around a wooden circle would be difficult for men without metal tools, but the placing of only 56 holes would be less difficult. Let us suppose it to be done. The possibility of making up a pegboard to represent the Aubrey Circle then raises an inverse question: why bother to construct the Aubrey Circle at Stonehenge at all? Why not simply keep a real-time account of the positions of $S$, $M$, $N$, $N'$ on one pegboard, and use a second, similar pegboard for making predictions? Such a procedure would be possible, but risky, since a pegboard would always be vulnerable to an unintentional displacement of the pegs. No doubt with due precautions we could succeed in operating such a system for a while, but sooner or later an accident would happen, and we would then have to wait several years before the nodal points could be reset. After one such unfortunate experience, we would almost surely decide to follow the builders of Stonehenge, by keeping the Aubrey Circle at Stonehenge as an inviolate reference standard. With heavy stones there, representing markers for the Sun, Moon, and nodal points, accidental displacements could not happen. A mess-up

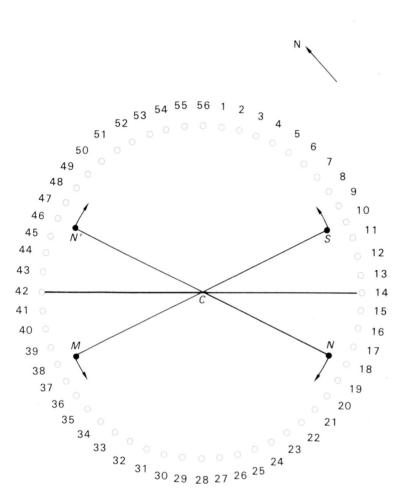

Figure 5.1. A wooden pegboard with 56 equally spaced holes around a circle. The pegs *S*, *M*, *N*, and *N'* are used to represent the positions of the Sun, the Moon, and the nodes of the Moon's orbit.

of the real-time positions on our pegboard could then easily be corrected by a visit to the site itself.

The use of a pegboard of manageable size, say, 6 feet across, overcomes the need to shout instructions across the Aubrey Circle at Stonehenge. Indeed, a pegboard could be operated by just one or two persons. Starting with pegs for $S$, $M$, $N$, $N'$ in the current real-time positions, let us imagine ourselves to go ahead in steps of half a day, recording each step by threading a bead on a piece of string. At each such step we move the peg for $M$ by one hole counterclockwise, and after 13 half-days we move the peg for $S$ by one hole, also counterclockwise. Keeping a running count of half-days up to 13 in one's head would be possible, but I myself would prefer to be relieved of this necessity. One recourse would be to refer to the tally, perhaps by arranging the beads in blocks of 13, which one could do by using a special colour for every thirteenth bead.

The pegs for $N$ and $N'$ are to be moved clockwise by three holes a year. Instead of doing this by referring to the tally, it would be simpler to refer to $S$. As $S$ goes through one-third of a full circuit of the board, move $N$ and $N'$ by one hole. For example, suppose we happen to start with $S$ at hole 56, and with $N$ and $N'$ at holes 21 and 49, respectively, the holes being numbered in accordance with the archaeological convention shown in Figure 5.1. Because the holes are numbered clockwise, both $S$ and $M$ move backward with respect to the numbers. We proceed to move $M$ and $S$ as described in the preceding paragraph; when $S$ reaches hole 38, we move $N$ to hole 22 and $N'$ to hole 50. We continue further until $S$ reaches hole 19, whereupon $N$ and $N'$ are moved to holes 23 and 51, respectively. And when $S$ arrives back at hole 56, $N$ and $N'$ are moved for the third time, to holes 24 and 52. A similar procedure is followed for further circuits of $S$.

Although slightly complex, this practical procedure for moving $S$, $M$, $N$, $N'$, could be mastered with a little practice – it would hardly be more difficult than learning to operate a camera. Instead of painstakingly working through each half-day separately, one would soon be able to move ahead in blocks of 13

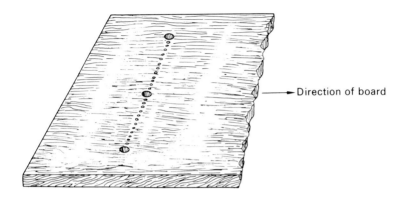

Figure 5.2. One end of a long rectangular board, with three
Aubrey holes represented (by the large holes), and with 13
subdivisions between one Aubrey hole and the next.

half-days. A complete circuit of *S*, representing the passage of
a year, could readily be completed in an hour or two of shuffling
the pegs.

Once a suspected eclipse was reached, with *S* and *M* close
to the nodal points, the above-mentioned difficulty (2) would
arise. A finer control of the markers would then be needed, such
as might be achieved with the aid of an auxiliary board shaped
like a long rectangle, as in Figure 5.2. Only three of the Aubrey
holes are marked at each end of this auxiliary board, which,
because of its length, gives an expanded scale for the separation
of neighbouring holes. The expanded scale provides space for
subdivisions to be inserted between adjacent Aubrey holes. With
13 subdivisions, a Sun marker moved by one subdivision each
half-day would keep track of the Sun to within 0.5°. The use
of 13 subdivisions happens also to be convenient for giving a
fine control of *N* and *N'*. Since there are approximately 13 lunar
circuits in a year, the rule for moving *N* and *N'* on the auxiliary
board is exactly like that for the main board, except that the
Moon replaces the Sun. On the main board *N* and *N'* are moved
three Aubrey holes *each year*. On the auxiliary board, *N* and
*N'* are moved three subdivisions each *lunar month*.

The positioning of $S$ relative to $N$ and $N'$ could thus be refined on the auxiliary board, whenever a suspected eclipse configuration was encountered on the main board. Some care would be needed to work this system, but the method permits repetitions and cross-checking by several operators, and meanwhile the reference standard back at Stonehenge is never lost.

Now that we have seen how to operate the rules of Chapter 3 in a practical way, we can pass on to consider how a crucial improvement must be made. The rule for moving $M$ on the main pegboard, one hole each half-day, is the least accurate of the three rules of Chapter 3. After only one circuit in 28 days of the pegboard, the marker $M$ would deviate from the true position of the Moon by more than one Aubrey hole. Such an error need not be serious for the reference standard at Stonehenge itself, because the position of $M$ can be reset there at every full moon and at every new moon (better, I would think, at new moon). No resetting is possible, however, during the making of predictions. If we seek to predict ahead for a whole year, there will be some 13 circuits of $M$ on the main pegboard, and in these circuits $M$ will go out of position by as many as 18 of the 56 holes. This would destroy all hope of accurate predictions.

An improved rule for moving $M$ is obviously needed. Consider the effect of moving $M$ as before, one hole each half-day, but with the addendum that on every circuit of the main pegboard a certain hole is always skipped. The time around the main board is then 27.5 days, closer to the actual period. Suppose, further, that once in every third circuit a second hole is skipped. The time around the board now becomes 27.333 days, very little different from the actual period of 27.322 days. With this modified rule, the position of $M$ can be predicted with adequate accuracy a year or two ahead of time.

It is interesting to ask if this addendum for moving $M$ was applied at Stonehenge itself. If so, we might expect two of the 56 Aubrey holes to have been unmistakably marked. In fact, two of them were, holes 19 and 44, lying in the ditches which surround Station positions 92 and 94. These ditches are marked on the official ground plan, and in Figure 2.3, while that at 92

Figure 5.3. The ditch at Station position 92 is clearly seen at lower left; that at Station position 94 is immediately to the right of the entry point of the visitor's path. (Courtesy of Aerofilms and Aero Pictorial, Ltd.)

can be seen in Figure 5.3. Suppose we elect to skip hole 44 in every circuit of the Moon marker. Then in the other ditch at mound 92 we set up an auxiliary marker stone which can rest in three possible positions, at hole 19 itself, at a position defined to be clearly outside the Aubrey Circle, and at a position inside the Aubrey Circle. The auxiliary marker is to be moved, say, in a counterclockwise sense, from one of these positions to the next whenever $M$ skips over hole 44. There are now two possibilities as $M$ approaches hole 19. If the auxiliary marker is occupying 19, that hole too is skipped by $M$, but if 19 is empty, then $M$ moves into it in the usual way. This procedure gives the average period of 27.333 days for $M$. By using the actual arrangement of Stonehenge, one could thus obtain essentially the correct period for $M$ without needing any additional equipment, except for a stone to serve as an auxiliary marker.

I come now to an entirely new form of error, one that would almost certainly be beyond the capability of men without metal instruments. The Moon's orbit around the Earth is not strictly a circle. The slight ellipticity of the actual orbit causes the Moon to move faster over some parts of its path on the sky than over other parts. Astronomical history tells us that the problem of allowing for this non-uniformity in a predictable way (to the first order for the eccentricity of the Moon's orbit) was solved by the Greek astronomer Hipparchus around the year 130 B.C. Fortunately, this source of error, unavoidable in the third millennium B.C., is not cumulative – it does not add from one circuit of the Moon to the next. In fact, the effect shows up after the passage of only a few days, and then never gets worse. The error in the pegboard system arising from this cause would be about one Aubrey hole in the position of $M$. That is, at some places in its monthly circuit, the Moon marker $M$ would be out of position, compared to the true position of the Moon, by about one Aubrey hole. This unavoidable source of error has at least the positive merit of making it unnecessary to bother with smoothing the jumps of $M$ at holes 19 and 44.

Since $S$, $N$, $N'$ move slowly compared to $M$, an error of one Aubrey hole, corresponding to a timing error of half a day in the position of the Moon, scarcely affects at all the geometrical sequence of configurations of $S$, $M$, $N$, and $N'$. Consequently, the error does not affect the question of whether an eclipse will occur in a particular circuit of $M$. The effect of the ellipticity of the Moon's orbit is to introduce an error of half a day into the time when the eclipse will occur. At first sight this imprecision might not seem too important, but it would in fact be a source of considerable irritation. It would destroy the ability of the pegboard operators to decide whether the British Isles would be turned towards the Moon at the moment of eclipse, and so would prevent them from predicting whether the eclipse would be seen or not.

Yet this remaining problem could be solved by making another observation. To recapitulate, lunar eclipses occur when the Moon enters the shadows cast by the Earth. The shadow is effectively a cylinder with a radius of about 4,000 miles, about a sixtieth of the radius of the Moon's orbit. The centres of the Sun, Earth, and Moon must therefore be aligned during a lunar eclipse to within an angle of not more than $1°$. A similar alignment is also necessary for a solar eclipse. Hence the Moon must lie close to the Earth–Sun line at the time of an eclipse, the Moon being 'full' near lunar eclipse and being 'new' at solar eclipse.

Now, the future position of the Earth–Sun line is known with better accuracy from our pegboard than is the future position of the Earth–Moon line, even though an error of the same kind as that which affects the lunar motion is also present for the Sun. Because the Earth's actual orbit around the Sun is not quite circular, the Sun appears to speed up over certain sections of its path and then to lag over other sections. This lack of uniformity in the apparent motion of the Sun on the sky would produce an error, between the position of the marker $S$ and the true position of the Sun, amounting to about 1/3 of a division on the Aubrey Circle. Thinking of the Aubrey Circle as the Sun's path among the stars, we would know the background of stars against which a predicted eclipse will take place to within about $360/56 \times 1/3°$, i.e., about $2°$. So this error for the Sun is not as large as that for the Moon.

What can next be done, still some months ahead of an eclipse, is to observe the Moon from day to day. When the Moon reaches the appropriate background of stars, a known number of complete circuits, a known multiple of 27.32 days, must bring us forward in time to the moment of the eclipse. In this way the timing of an eclipse could be calculated to within the accuracy for the future position of the Earth–Sun line, i.e., to within 1/3 of a division on the Aubrey Circle, which the Moon covers in 1/6 of a day. In most cases this timing would be accurate enough for us to know whether an eclipse would be seen or not. Moreover, as the moment of eclipse approached, the inaccuracy of 1/3 of a division on the Aubrey Circle could be reduced substantially, by actually watching the Sun in its relation to the stars. The question of whether an eclipse would be seen or not could therefore almost always be decided by, say, a month before its occurrence. Only a few marginal cases would remain, with the eclipse occurring just above or just below the horizon; and since eclipses take an hour or so to develop, the Moon would be likely to rise at some stage in these marginal cases, making it safe to predict that at least a portion of the duration of such eclipses would be seen.

We have now worked through all the problems involved in arriving at successful eclipse predictions. No equipment needs to be added to the structure of Stonehenge I, except marker stones for $S$, $M$, $N$, $N'$, and a stone to serve as a counter at position 92. Nor is any auxiliary equipment required that would go beyond what might reasonably have been available in the third millennium B.C.

So far, I have said very little about the actual moment of an eclipse. The most striking of all eclipses is the total blotting out of the Sun by the Moon. But total eclipses of the Sun, in any one geographical location, are too rare a phenomenon for them to have played a large role in the daily lives of ancient peoples. Only two solar eclipses in the twentieth century (1927 and 1999) have zones of totality in the British Isles. It seems unlikely, therefore, that the primary goal of the builders of Stonehenge was to predict total eclipses of the Sun. Nor could they have achieved this goal, for the prediction of the geographical zones of totality for such eclipses requires intricate observations and calculations of an entirely modern kind.

Partial eclipses of the Sun, although much more frequent, are not very noticeable. Of the several scores of partial solar eclipses which have occurred in my own lifetime, I doubt if I have noticed more than three or four. The eye makes automatic compensation for variations of the light intensity, so that, unless most of the Sun is obscured by the Moon, or unless we are deliberately looking for them, partial eclipses of the Sun easily pass us by. The builders of Stonehenge might very well have been concerned to look deliberately for partial eclipses of the Sun, but it seems rather unlikely that this could have been their primary aim.

Eclipses of the Moon are quite another matter. They occur only on nights when the Moon is full. Because the Earth is much larger than the Moon, the Earth is highly effective in blocking sunlight from the Moon, so that lunar eclipses are quite frequently total. It would therefore be eclipses of the Moon for which the builders of Stonehenge would mostly have been looking. Since there are one or two such eclipses per year, prediction for about a year ahead of time would be required for them, just as we have so far considered.

I would suppose that the builders of Stonehenge believed the Sun and the Moon to be divinities whose eclipse was a phenomenon striking at the very foundation of the universe. Although total eclipses of the Sun could not be predicted year by year, their rare occurrence would add a sensational impact to them. The Sun might well have been thought masculine, a god; the Moon feminine, a goddess – an association suggested by the correspondence between the period of the phases of the Moon and of the menstrual cycle of women. The Sun and Moon together would thus be a divine representation of man and woman. On eclipse, the Moon, and very occasionally the Sun, dies. But they are then immediately and miraculously reborn, as many people believe even to this day that we ourselves are.

These, I would suppose, were the beliefs held by the builders of Stonehenge. No doubt the marker stones $S$, $M$, $N$, $N'$, came to acquire the qualities of the divinities themselves, and so became holy objects, a transference of a god-like quality to an inanimate object which still permeates modern religions. The moving of the marker stones would then have been a matter of ritual as well as of science. The goddess, represented by $M$, is required always to jump across Aubrey hole 44 in the ditch associated with Station position 94. There is also a conditional jump of $M$ at hole 19, depending on an auxiliary marker that moves around the ditch at position 92. There are three positions for the auxiliary marker – and, as if to remind us of this, there are three Aubrey holes in the mound and ditch at position 92.

Can an interpretation also be found for the four Aubrey holes 44 to 47 within the ditch at position 94? The Sun marker $S$ moves each year into this portion of the Aubrey Circle and remains there for 26 days, close to the 27.3-day lunar period. From this near-equality, one can see immediately that the fast-moving Moon will usually arrive during the 26 days in this same territory, and so will reside there for a while together with the Sun. Interestingly, our rules for moving $M$ and $S$ always lead to an annual association of $M$ and $S$ within the ditch at position 94, although there is a limiting case in which the association is of the smallest possible duration, a veritable brief encounter.

To understand this limiting case, let us make the rule that whenever $M$ and $S$ have both to be moved on the same morning, or in the same evening, one or the other is always moved first, say, $M$. The limiting case then occurs when $M$ hops out across hole 44 just before $S$ arrives at hole 47. After a further 25.5 days, $S$ will be at hole 44, with one half-day still to remain there. Meanwhile $M$ will have reached hole 48. After another half-day, the Moon, moving first, goes within the ditch to hole 47 just before $S$ goes out to hole 43. Brief contact is thus established.

This junction of $M$ and $S$ within the specially marked enclosure at position 94 may well have been thought to be of profound significance, perhaps being connected in the minds of the people with the death and rebirth of the Moon at a subsequent eclipse – and so with the life and death of the human species itself. It is ironic, and perhaps not a little sacrilegious, that in modern times the visitors' path enters Stonehenge across this specially marked territory.

# 6

# *The Cultural Connections*
# *of Stonehenge*

*The* astronomer, examining the evidence, can hardly fail to be impressed by the story of Stonehenge. In my own case, whatever prejudices I had in approaching the subject were directed against the claims of Professor Hawkins and Mr. Newham, but I soon saw that the 'anti-view' simply could not be maintained. In this respect, Stonehenge differs from other ancient monuments. Astronomical alignments can certainly be found in other monuments, but, with the possible exception of Callanish in the Hebrides, they appeared to me to be without the deeply laid purpose of Stonehenge.

Yet if the eclipse interpretation of Stonehenge be admitted, the intellectual achievements of the builders must stand out in prehistory like a veritable Mount Everest. And there must surely be forerunners to Stonehenge. Where are they? And why has no account of the achievement come forward to us in the full light of documented history? These questions demand answers. Since satisfactory answers have not been given to them, the historian and prehistorian maintain a position of sceptical doubt. Indeed one of the guidebooks available for purchase at Stonehenge goes so far as to remark about the eclipse interpretation: 'These claims are so extravagant that they cannot be taken seriously.'

In any impasse between distinct disciplines, it is inevitable that each side will seek to weaken the position of the other side. An argument used by prehistorians, one with which I can feel some sympathy, is that, in developing the astronomical interpretation of Stonehenge, the astronomer is merely impressing his own state of mind on the monument. But if this were true, the astronomer should also be able to impress his state of mind on monuments other than Stonehenge. Following what I felt to be exciting results coming from the eclipse interpretation of Stonehenge, I tried (quite hard) to apply similar ideas to Avebury. The result was a failure. For me, therefore, the argument is not true. The extent to which I am able to impress myself on other monuments, although not entirely zero, has a characteristic and recognizable weakness about it, contrasting most strikingly with the case of Stonehenge. If the structure of Stonehenge, particularly of Stonehenge I, had not suggested it, I doubt if any modern astronomer would have conceived that eclipses could have been predicted by men equipped only with sticks, stones, ropes, and primitive tools for the shaping of wood.

In this chapter I shall attempt to answer the historians and prehistorians by seeking to weaken their position. Let me start, however, from what I take to be a point of agreement. Stonehenge, both in its intellectual concepts and as a physical structure, could not have been an act of special creation. Although megalithic monuments are common in the British Isles, and in northwestern Europe generally, not one that I have seen looks to me like a forerunner of Stonehenge I. Some may have a connection with astronomy, but more, it seems, as a distant memory of times gone by. I see these other monuments as belonging mainly to the second millennium B.C., about a thousand years later than Stonehenge I. By then, I suspect, astronomy had largely been replaced by ritual. Even Stonehenge III appears to me to be more concerned with ritual than with astronomy. Stonehenge I is essentially very simple, a set of marked positions and a few naturally occurring boulders – there may even have been wooden posts instead of boulders in the beginning. This simple structure was sufficient for the astronomical needs. Once simplicity became replaced by complexity, as in Stonehenge III, one can be virtually certain that science had been displaced by ritual.

In seeking for the forerunners of Stonehenge I, we must therefore look for something simple, we must look around the year 3000 B.C., not around 2000 B.C., and we must look in the right place – which is where? Some years ago, I attempted to argue that the indigenous population of the British Isles might have developed the astronomical ideas which seem to have motivated the construction of Stonehenge I. I am now inclined to accept the objection that such an intellectual achievement was beyond the capacity of the local neolithic farmers and herdsmen. The sharp latitudinal dependence of the Sun–Moon alignments of the Station-position rectangle suggests that the builders of Stonehenge I might have come into the British Isles from outside, purposely looking for a site where this rectangular alignment would be possible, just as the modern astronomer often searches far from home for places to build his telescopes. Stonehenge I would then have been an act of special creation for the British Isles. The forerunners of Stonehenge I would have to be sought elsewhere, and the search is then broadened so greatly that I feel no surprise, or unease, that the forerunners are so far unknown.

Commenting on someone else's field of study is always a hazardous business. Nevertheless, I would not be stating my position clearly if I avoided mentioning a specific objection which I have to the methods usually followed by the prehistorian. Of necessity, the archaeologist must be concerned with artifacts that have survived long periods of time. Tools, pottery, and mortuary relics form the backbone of archaeological studies. The danger arising out of this necessity is that the archaeologist will come to believe that what he can know about – the making of tools and pots, and the burying of the dead – is also all there was to the life of prehistoric peoples. It is all too easy for the views of the prehistorian to become biased by the survival characteristics of the various objects available for study. Since the intellectual conceptions and abstractions of a non-literate people rarely had any way to survive, the tendency is to suppose there were none, which is why the discovery of the great Stone Age cave paintings in Spain and France came as an intense surprise to the prehistorians of the nineteenth century.

The intellectual activity of mankind during prehistory is a vast, almost uncharted ocean. Truly vast. I well remember the shock with which I first realized that more than ten billion people have lived in prehistoric ages, probably many more than have lived in classical and modern times. The familiar curve showing how the world population has grown with time is thus misleading, since it suggests that only a few miserable souls managed to eke out a living in the days before written history. There have been only about 200 generations of history. There were upwards of 10,000 generations of prehistory. Most of the people who have ever lived, lived then, not now. Among the great throng, it seems to me likely that some must have gazed up at the sky and wondered earnestly about the Sun, Moon, and the stars. They would have done so with a basic intelligence equal to our own.

There is an analogy between Stonehenge and the old evolutionary problem of the electric eel. The eel lives by stunning or killing its prey by means of the powerful shock it can deliver. In a Darwinian sense, one can see how present-day eels might evolve into future eels that can produce still more powerful shocks, but how did the electric eel arrive at this way of life in the first place? A feeble shock, such as the early forerunners of the present-day eel must have had, would not have stunned anything, and so would have been useless for securing food. The answer to this problem is that the electric properties of the early forerunners of the present-day eel were used in a different way, probably for navigation. In a like manner, eclipse prediction is far too developed and complex a concept to have existed in the beginning. The early astronomical ideas must have been used to solve less difficult problems. Just as the archaeologist will not be satisfied until forerunners of the actual structure of Stonehenge have been traced, so it is necessary to trace the forerunners of the astronomical ideas themselves. It must have been possible to go from simple beginnings in a sequence of mounting complexity until the ideas necessary for eclipse prediction came at last into the perceptions of the early watchers of the sky. Let us consider how this might have happened.

First, let me clear out of the way the problem of apparently esoteric pieces of information, which the early observers could actually have acquired without too much trouble. The eclipse interpretation of Stonehenge requires that the builders knew, with reasonable accuracy, that the Moon's orbital period is 27.32 days, that the length of the year is 365.25 days, and that $N$ and $N'$ have an 18.61-year period. What method could have been used to arrive at these rather precise numbers?

The path of the Moon can easily be followed among the stars. Starting from a particular constellation of stars, one could judge quite accurately when the Moon next reached the same configuration of stars, perhaps to within an angular tolerance of $1°$. Without a clock, however, the time required for such a circuit could not be measured with any comparable accuracy. Suppose we content ourselves to measure time in units of half a day, simply by counting the number of light and dark intervals. Then we shall be able to say that the time required for a complete circuit of the Moon is more than 54 half-days but less than 55 half-days. For a single circuit we would thus have an error of about 2 per cent. But suppose that, instead of observing a single circuit, we observe 1,000 circuits. Since the errors of position and of timing will be no worse than before, the measurement becomes accurate to within 0.002 per cent. In this very simple way, we would know that the Moon's period is 27.322 days, and not 27.321 or 27.323 days.

A similar idea would serve to measure both the length of the year and the 18.61-year period of the nodes, although in the beginning the prehistoric observers would hardly be concerned with the nodes. In the beginning the observers would be concerned only with visible objects, particularly with the Sun and Moon. After a few hundred circuits of the Sun among the stars, it would be known to within an accuracy of about 0.01 per cent that the sidereal year had 365.25 days.

What interests would drive the observers to make these measurements? The apparent relation of the period of the phases of the Moon to the menstrual cycle of women would have raised the question of whether this seeming connection was real or not. The Moon would be carefully observed, and scientifically minded persons would be led to a careful measurement of the lunar orbital period. Observations of the Sun as it passed through the star constellations of the Zodiac would show the motion of the Sun to be apparently associated with the seasons of the year. Was the connection real or not? This question would call for a precise measurement of the length of the year, of the period of the Sun's motion in the sky, as well as for a good way to measure the period of the seasons. The solution to this latter problem would be to observe the yearly swing of the rising Sun. When it was found that the two periods really did seem to be the same, the measurements would be pressed to tighter and tighter margins. Always the two periods would seem to be the same.* This would be a great success, an exciting and encouraging discovery. A connection would have been established between life on Earth, as expressed by the seasons, and what was happening in the sky.

At an early stage of these observations, it would be noticed that, although the Moon's path in the sky was very like the Sun's path, the two were not quite the same. As the work proceeded from year to year, it would then be found that the places where the Moon's path crossed the Sun's path were slowly changing. Many eclipses would occur during the century or two required to establish the Sun's period with great accuracy, and the astonishing fact would gradually emerge that eclipses never occur unless both the Sun and the Moon are close to one or the other of the two places where their paths cross. This remarkable discovery would intensify the observational interest in the nodal points. When, from continuing observation, the nodes were found to move along the Zodiac in a quite regular way, completing a circuit in 18.61 years, the concept of eclipse prediction would be possible. For if the Sun, the Moon, and the crossing points of their paths all move in a smooth way in known durations of time, why should their positions not be forecast ahead of time?

---

*To the modern observer, there is a slight difference: the motion of the Sun among the stars gives the sidereal year, and the swing of the rising Sun gives the tropical year. But the difference is only one part in 26,000, and this would probably lie beyond the capability of neolithic man.

This brief synopsis, of a development from simple beginnings to the concept of eclipse prediction, assumes the forerunners of the builders of Stonehenge to have been numerate. They would have needed to keep tallies of half-days, even when the number of half-days became large; there are 13,595 half-days in 18.61 years. A tally of so large a number could not have been kept by memory; it would require physical storage in an external form. Without paper and writing, the natural storage would be by wooden beads on a string, but since it hardly seems likely that 13,595 separate beads would be mounted in this way, the concept of orders of numbers would have been needed.* The orders need not have been tens, hundreds, thousands ... although the ten fingers of the two hands make this a natural choice. The different orders could conveniently be represented by differently coloured beads mounted on different strings.

The concept of arithmetic division would also be needed, since the tally of half-days would require division by the number of circuits of the Moon, or of the Sun, or of the nodes in their motion around the Zodiac. If the number of circuits were suitably chosen in relation to the order of the numbers, division would be simple. For example, if the decimal scale were used, division by 100 circuits requires simply a twofold decrease of the orders of the numbers appearing in the tally. This method of division would force the concept of a fractional part of a day, or a fractional part of the year – the fraction is the bit left over when the orders of the numbers are reduced in this way.

The eclipse interpretation requires an apparently subtle piece of astronomical information: that in the month of the 18.61-year cycle when the Moon swings most to the north, the node $N$ lies behind the midsummer position of $S$ by 90°. But this information is subtle only if one is required to understand it! No subtlety is required to arrive at this fact by direct observation. I would suppose the forerunners of the builders of Stonehenge, following the motion of the nodes along the Zodiac, observed

---

*An actual attempt to store 13,595 beads would surely force one to invent the concept of orders of numbers!

how their position was related to the Sun and to the monthly swing of the Moon. Certainly in the many circuits of the nodes required to determine the 18.61-year period, they would have had ample opportunity to do so.

The earliest observers would almost surely have divided the Sun's path according to the bright stars which happen to lie along the Zodiac. It is a big step conceptually to pass from such a division – perforce into rather unequal steps – to a division into equal steps. The latter requires abstract points of division, an abstract division that was given representation at Stonehenge by the Aubrey holes. It was also a big step, perhaps the biggest step of all, to complete the full circle of the Zodiac. In midwinter, an observer can see most of the constellations of the Zodiac, but a few remain hidden by the bulk of the Earth.* The visible constellations lie along a circle, and as the night hours pass, further constellations rise in the east, as previously visible ones set in the west. What the neolithic observers had to do was to complete the partially observed circle *in their minds* and then to lay it out on the ground in the form of the Aubrey Circle. Provided this step was made, nothing else stood in the way of determined and interested observers.

So far we have concerned ourselves only with the problem of the forerunners of Stonehenge. We must turn now to the question of why the achievement of Stonehenge did not become a major cultural pattern of written history. The answer, I believe, is that the method of eclipse prediction used at Stonehenge became replaced by a second, quite different method, one that was an outcome of the invention of writing. Let us consider in some detail how this further development could have come about.

Let us examine what happens to the positions of $S$, $M$, $N$, $N'$ in a time of 6,585 days, the positions being taken with respect to the background of the stars. There are 18 complete circuits of $S$ around the Zodiac, requiring $18 \times 365.25 = 6574.5$ days, and

---

*In midwinter, travellers to the north of Scotland would see almost the full circle.

a bit of an extra circuit which occupies 10.5 days. In this 10.5 days, the Sun moves by $360° \times 10.5/365.25 = 10.35°$. Thus in 6,585 days the Sun overshoots 18 complete circuits by rather more than $10°$ counterclockwise from its starting point.

We know that $N$, $N'$ complete a circuit of the Aubrey Circle in 18.61 years, which is 18 years plus $0.61 \times 365.25 = 223$ days. It follows that after 18 years and 10.5 days, i.e., 6,585 days, the points $N$, $N'$ will still be short of a complete circuit, by $223 - 10.5 = 212.5$ days. In angular terms, $N$, $N'$ will be short of a complete circuit by

$$\frac{360 \times 212.5}{18.61 \times 365.25} = 11.3°.$$

Since $N$, $N'$ move clockwise on the Aubrey Circle, the angle $11.3°$ required to complete the circuit of $N$, $N'$ evidently overlaps the angle of $10.3°$ by which $S$ has overshot its 18 completed circuits. Hence, after 6,585 days, the relative relationship of $S$, $N$, $N'$ is the same as it was to start with, to within a tolerance of about $1°$.

To find where the Moon has moved to after 6,585 days, we need to know the lunar orbital period rather accurately, probably more accurately than it was known in ancient times; it is 27.3217 days. It takes $27.3217 \times 241 = 6584.53$ days for the Moon to complete 241 circuits, with 0.47 days still to go. In this 0.47 days, the Moon will move counterclockwise through an extra angle of

$$\frac{360 \times 0.47}{27.3217} = 6.2°.$$

Thus the Moon will still have about $4°$ to go before the Earth–Moon line swings into the same relation to the Earth–Sun line that it had in the beginning. Because the Moon moves so quickly compared to $S$, $N$, $N'$, the latter scarcely move at all as the Moon covers this $4°$, which it does in only $4/360 \times 27.3217 = 0.30$ days. Hence in 6585.3 days, the relative relationship of $S$, $M$, $N$, $N'$ is very close to being the same as it was in the beginning. The relationship of $S$, $M$, $N$, $N'$ has simply swung round bodily through an angle between $10°$ and $11°$ counterclockwise from its starting point as measured against the background of the stars.

*It follows that 6585.3 days after an eclipse, there will be another eclipse of the same kind.*

Equipped with this knowledge, we now have a conceptually simple method of predicting eclipses. Whenever an eclipse actually occurs, start a tally of days. When the tally reaches 6,585, an eclipse will occur approximately one-third of a day later than the first eclipse. The only awkward feature of this method is that, because of the 0.3-day delay, an eclipse might not be seen on the second occasion. But by taking note of the time of day or night when the first eclipse occurred, we would know whether the second eclipse would be seen or not. Once the system had been in operation for three or four cycles of 6585.3 days, we would be able to predict essentially every eclipse, and whether it would be seen or not.

Now we must ask for the line of evolutionary thought that could have led to the discovery of this second method of eclipse prediction. Notice that there is no rationale that could have led to the discovery. The fact that after 6585.3 days the angular configuration of $S$, $M$, $N$, $N'$ is repeated to within a fraction of a degree is a pure fluke. If the radius of the Earth's orbit around the Sun, or of the Moon's orbit around the Earth, had been a little different, the fluke would not have existed.

If the pegboard operators of Stonehenge could have gone ahead accurately in time by 18 circuits of the Sun marker the fluke would easily have been discovered. But the need to work operationally in terms of only 56 Aubrey holes introduced an error of $1°$ for each circuit of $S$. Prediction ahead for one or two years was possible, but not for as many as 18 circuits of $S$.

In my view there was no possible way to discover the fluke of this 6585.3-day eclipse cycle, known as the *Saros*, until after the invention of writing. It would scarcely be possible to collate happenings during at least two or three of these cycles of 18 years and 11 days without a recourse to written records. The process of discovery is more complex for written records than it is for keeping of a numerical tally, such as was required to discover the 18.61-year period of $N$, $N'$, since the observers knew beforehand that the latter cycle existed; they needed only to measure it. But the existence of the cycle of 18 years and 11 days could not have been known beforehand – it would need to be dug *a posteriori* out of the records.

Even with day-by-day records, it would still be difficult to recognize a cycle of 6,585 days, and I doubt the discovery could have come in this direct way. However, there is a way of keeping records, other than day-by-day, which would have made the discovery almost unavoidable. At the first eclipse of the cycle, the Moon must either be full (lunar eclipse) or new (solar eclipse), and at the second eclipse the Moon must also be full or new, in the same configuration as before. Between the beginning and end of the cycle, there must thus be a definite number of periods of the phases of the Moon. The time from full moon to full moon (or from new moon to new moon) is 29.5306 days. [When the Moon has completed a full circuit, in 27.32 days, with respect to the background of stars, it still has to go on for an extra 2.21 days to catch up the moving Sun, since it is the relation of the Earth–Moon line to the Earth–Sun line which determines the occurrence of full moon (or new moon), not the relation of the Earth–Moon line to the stars.] The time required for 223 such periods of the phases of the Moon is $223 \times 29.5306 = 6585.3$ days. There are thus 223 *lunations*, as the period of the phases of the Moon is called, in the 6585.3-day cycle.

Considering that lunar eclipses occur only when the Moon is full, let us suppose that a chronicler decides to keep a record of happenings at the times of full moon. Within less than a century after the beginning of such a chronicle, the necessary data for the discovery of the 223-lunation eclipse cycle would have been accumulated. An assiduous researcher of past records would inevitably discover the cycle.

Suddenly the need to move the markers $S$, $M$, $N$, $N'$ painstakingly in the manner of Chapter 5 disappears. Instead of predicting ahead of time for only a year or two, we can now go ahead for several of our 6585.3-day cycles, for 50 years or more. The need to reset $S$, $M$, $N$, $N'$ by means of alignments at Stonehenge disappears. The actual occurrences of eclipses can themselves be used to calibrate the system. In fact, Stonehenge falls flat on its face, and this, I think, is why Stonehenge did not come through into the full light of documented history. Stonehenge was destroyed by fluke coincidences between the periods of the Sun, the Moon, and the lunar nodes. It was the Saros which came through into the light of history.

The line from the Earth to the Sun sweeps through the line connecting the Earth to the node $N$ in rather less than a year, as it does through the line from the Earth to $N'$. The time between successive coincidences of the Earth–Sun line with these nodal directions averages 346.62 days, less than a year, because the Sun and the nodal points move in opposite senses around the Aubrey Circle. The time required for a sequence of 19 such coincidences is therefore $19 \times 346.62 = 6585.8$ days. Thus the eclipse cycle of the Saros corresponds very closely to 19 complete circuits of $S$ with respect to the line connecting $N$ to $N'$.

Let us return to Stonehenge III, remembering the 19 blue-stones of the inner horseshoe. Did the use of the number 19 indicate a knowledge of the Saros? If so, the ideas of Stonehenge I must already have been superseded by the time of Stonehenge III. For applying the eclipse-prediction method of Chapter 5, Stonehenge III is much inferior to Stonehenge I, but, then, it could afford to be, if the builders of Stonehenge III knew of the Saros. The builders of Stonehenge III could afford to be all things to all men, paying respect to the tradition of the older culture of Stonehenge I, but putting their main effort into impressive ritual, secure in the knowledge of an easy and fool-proof method for the prediction of eclipses of the Sun and the Moon.

There is another interpretation, known as the *Metonic cycle*, which might be given for the number 19. In 19 years there are $19 \times 365.25 = 6939.75$ days, and in 235 lunations there are $235 \times 29.5306 = 6939.69$ days. Thus 19 years after the occurrence of a full moon, there will be another full moon. But would the prediction of a full moon 19 years ahead of time really seem an impressive performance? Since there is a full moon in every month, I doubt it.

On the other hand, this interpretation is explicitly mentioned in a passage quoted by Professor Hawkins in *Stonehenge Decoded* from the Sicilian historian Diodorus (*ca.* 44 B.C.); who is here quoting an earlier historian, Hecataeus (*ca.* 500 B.C.):

> This island ... is situated in the north, and is inhabited by the Hyperboreans, who are called by that name because their home is beyond the point whence the north wind [Boreas] blows; and the land is both fertile and productive of every crop, and since it has an unusually temperate climate it produces two harvests each year. Moreover, the following legend is told concerning it: Leto [mother of Apollo and Artemis–Zeus was their father] was born on this island, and for that reason Apollo is honoured among them above all other gods; and the inhabitants are looked upon as priests of Apollo, after a manner, since daily they praise this god continuously in song and honour him exceedingly. And there is also on the island both a magnificent sacred precinct of Apollo and a notable temple which is adorned with many votive offerings and is spherical in shape. Furthermore, a city is there which is sacred to this god, and the majority of its inhabitants are players on the cithara; and these continually play on this instrument in the temple and sing hymns of praise to the god, glorifying his deeds.
>
> The Hyperboreans also have a language ... peculiar to them, and are most friendly disposed towards the Greeks, and especially towards the Athenians and the Delians, who have inherited this goodwill from most ancient times. The myth also relates that certain Greeks visited the Hyperboreans and left behind them there costly votive offerings bearing inscriptions in Greek letters. And in the same way Abaris, a Hyperborean, came to Greece in ancient times and renewed the good-will and kinship of his people to the Delians. They say also that the Moon, as viewed from this island, appears to be but a little distance from the Earth and to have upon it prominences, like those of the Earth, which are visible to the eye. The account is also given that the god visits the island every nineteen years, the period in which the return of the stars to the same place in the heavens is accomplished; and for this reason the nineteen-

year period is called by the Greeks the 'year of Meton.' At the time of this appearance of the god he both plays on the cithara and dances continuously the night through from the vernal equinox until the rising of the Pleiades, expressing in this manner his delight in his successes. And the kings of this city and the supervisors of the sacred precinct are called Boreades, since they are descendants of Boreas, and the succession to these positions is always kept in their family.*

The problem with this passage is to know what parts, if any, should be taken seriously. Lemprier's *Classical Dictionary* refers to Diodorus in the following terms: 'The author is too credulous in some of his narrations, and often wanders far from the truth.... He often dwells too long on fabulous reports.... His manner of reckoning by the Olympiads and the Roman consuls will be found to be very erroneous.' The reference to the Metonic cycle of 19 years is thus of doubtful value. I would suppose that Diodorus knew of this cycle, and that he simply mixed it with the fabulous report on the Hyperboreans. The report itself seems to have a legendary quality, with some general correspondence to Stonehenge itself – the island, the spherical (circular) temple, the mention of 19 years, all suggest a distant memory of a true situation. What I find hard to credit, however, is that a festival held only once in 19 years could have had much cultural significance or influence. To play an important part in the life of a community, festivals must surely be held at least once or twice each year.

The earliest festivals, probably predating Stonehenge by many thousands of years, would be held, I feel sure, in correspondence with the seasons of the year. The natural choices would be midsummer and midwinter, and if more were needed, quartering the year at the spring and autumnal equinoxes would be likely. With the coming of the ability to predict eclipses, such choices for festival dates would be convenient no longer, however, since the important eclipse events would not usually occur

---

*Diodorus Siculus, *The Library of History*, Book II, Chap. 47, as translated by C. H. Oldfather in the Loeb Classical Library (Harvard University Press and Wm. Heinemann, Ltd., 1935), II, 37–41.

on festival days. The sensible thing to do would then be to re-place the two equinoctial festivals by two eclipse festivals, still keeping the seasonal festivals of midsummer and midwinter. Eclipses can occur only when the Sun marker $S$ is close to the nodal line $N$ to $N'$. The Sun makes a complete circuit with re-spect to this line in the 346.62 days mentioned above, and in making a complete circuit, it crosses the line twice, at intervals of approximately 173 days. These occasions would form the basis for the fixing of eclipse festivals, but the precise dating of such festivals would depend on the eclipse predictions them-selves, which could vary from the regular 173-day interval by a few days either way.

In such a system we have two festivals determined by the 'eclipse year' of 346 days, some 19 days shorter than the seasonal year of 365 days. Hence the eclipse festivals would occur 19 days earlier each year relative to the seasons, and there would thus be times when the eclipse festivals would coincide with the seasonal festivals. Such a situation would occur approximately every ninth year, this being half of the 18.61-year period of the nodal line $N$ to $N'$. I can well believe that a falling together of the two kinds of festival would be treated as occasions for special rejoicing in the manner described by Diodorus. Remembering the discussion at the end of Chapter 3, we can suspect that the midwinter falling together of the festivals would have been the occasions when the Moon, on a day of its eclipse, rose over the Heelstone. It is these particular occasions which Professor Haw-kins emphasizes in *Stonehenge Decoded*. He suggests a relation to Diodorus' 19-year festival, but the falling together of the fes-tivals occurs twice, not once, in the 18.61-year period.

It is, of course, the eclipse year, not the seasonal year or the 18.61-year period, which has the powerful relation to the number 19, since an eclipse occurring at one moment will reoc-cur almost exactly 19 eclipse years in the future. Notice that, although 19 eclipse years of 346 days must be waited for the repetition of a particular eclipse configuration, we do not have to wait 19 such years before any eclipse occurs. Eclipses occur all the time, at a rate of two or three per year. The situation is that *each* eclipse configuration generates a 19-year sequence of eclipses. Each configuration has its own roster of eclipses. I would suppose that, by the time of Stonehenge III, the keeping of such rosters had become the main preoccupation of Diodorus' priestly guardians of the sacred temple. Much as we may doubt the factual accuracy of Diodorus, I must thank him for mention of singers and dancers. I have a vision of dancers appearing on the festival days, a vision very different from the grimmer mortuary reports of the archaeologist. May I suggest, since special arrangements must be made for Stonehenge on midsummer day (otherwise the crowds that gather there would be entirely out of hand), that the site be reserved on that day for Morris dancers?

But the 19-year eclipse-year cycle was a seriously regressive step. It worked – it was capable of working magnificently – but it led nowhere, because it was based on a mere fluke coincidence among the periodicities of the Sun, Moon, and nodes. It destroyed the need for meticulous observation. Stonehenge I would have worked just as well even if the fluke had not existed. But Stonehenge I and its concepts and ideas were gone by the second millennium B.C. The concept of constructing an instrument to observe the world was gone, and in that, much was lost.

In the light of history, especially of the history of the Near East, we see astronomy giving way to numerology. It seemed more profitable to shuffle numbers around on parchment than to observe the skies. True, this had some profit in the development of mathematics. And observation did not quite die away, but it was directed now into other channels, particularly to the movements of the planets. Since these movements were found too difficult to interpret in a sensible way, they were married to numerology, and astrology became their offspring.

At Stonehenge III, a strong connection with the Sun and the seasons of the year was still maintained. But the dominance of the Sun, in its relation to the year, became challenged elsewhere. Keeping track of the Sun in relation to the nodal line $N$ to $N'$ demanded a connection with the old observational techniques, and indeed Stonehenge III maintained this connection. But in other places a different way of following the 19-year eclipse cycle was adopted. It seemed simpler just to count the number of lunations: 223 of them determine the eclipse cycle.

Indeed, the eclipse cycle can be followed either way, as 19 eclipse years or 223 lunations. A people believing in the dominance of the Sun would be likely to adopt the first way. A people who adopted the second way would come to believe in the dominance of the Moon. The time of 29.5 days from full moon to full moon would seem to them to be the crucial period in the universe, and attempts would be made to force the seasons of the year into a lunar calendar. Such an attempt was in fact made, and it has come down to us in modern times as the curious division of the year into months, and in the method still used for fixing the date of the Easter festival (the first Sunday after the first full moon after the spring equinox).

No doubt the old method did not die without a struggle. The conflict might already have begun even before the construction of Stonehenge I. It could indeed have been this conflict which drove the builders of Stonehenge I from their homes, perhaps in the Near East or elsewhere, and which caused them to seek refuge in the British Isles. We saw in Chapter 5 that if one were seeking to operate the system of Stonehenge I, it would be sensible to count in blocks of 13. With the demise of the old method, this method of counting would surely give place to the more natural practice of counting in blocks of 10. Those who continued in blocks of 13 would be exposed to derision, and 13 would become the 'unlucky' number.

Yet I am convinced that something very important remained from Stonehenge I. What we may wonder, would a people, believing in the godlike qualities of $S$ and $M$, think of $N$ and $N'$? Since these points are always opposite each other in the Aubrey Circle, they would presumably be regarded as the same entity. From observation, the people would know that eclipses never occur unless $S$ and $M$ are both associated with the entity $N + N'$. Doubtless, then, the entity $N + N'$ would come to be thought of as a still more powerful god, an unseen god able to overcome the visible divinities. Could this be the origin of the concept of an invisible, all-powerful god? With the later demise of the importance of the Sun god, the godlike quality of $N + N'$ would seem to have remained, to become at last the invisible God of Isaiah. Could a distant memory of $S$, $M$, $N$ be the origin of the doctrine of the Trinity, the 'three in one, the one in three'? I suspect so. In Stonehenge I may well be many roots of our present-day culture.

# Appendix
# The Astronomical
# Alignments of the Sun
# and the Moon

## 1. The Seasons

The seasons of the year are caused by the tilt of the Earth's axis of
rotation to the plane of revolution around the Sun. On a time-scale
of a few centuries, the direction of the axis of rotation is essentially
fixed in space. So, as the line from Earth to Sun changes position during
the year, the axis changes its orientation with respect to the Sun. The
phenomenon has no connection with the actual rate of rotation of the
Earth. The seasons would occur at the same times in the year if the
Earth rotated in 5 hours or in 50 hours. The seasons must therefore

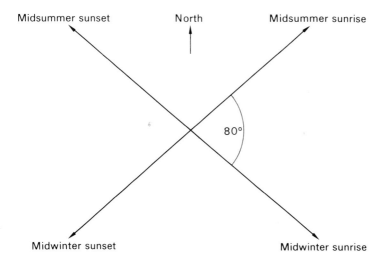

Figure A.1. The annual swing of sunrise and sunset, drawn for the
geographical latitude of Stonehenge.

be measured observationally by a method that is independent of the Earth's rotation rate. One such method, available to neolithic man, is illustrated in Figure A.1. For an observer situated at the centre of this figure, the direction of the rising sun varies from day to day. Between midsummer and midwinter there is a swing of about 80° at the latitude of Stonehenge. The seasons are thus correlated with the direction of the rising Sun, so that by observing the latter one can infer the former.

## 2. The Swing of Moonrise

The direction of moonrise also swings back and forth. It does so in a lunar month of 27.322 days. The angle of swing of moonrise is not the same, however, from month to month. At its largest, the angle is

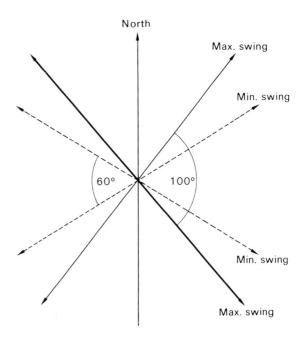

Figure A.2. Because of the changing angle between the plane of the Moon's orbit and the Earth's axis of spin, the monthly swing of moonrise (and moonset) varies from about 60° to 100° at the geographical latitude of Stonehenge.

about 100° at Stonehenge, and at its smallest about 60°, as illustrated in Figure A.2. Later we shall find that this effect arises because the plane of the Moon's orbit around the Earth is slightly inclined to the plane of the Earth's orbit around the Sun. The time required for the angle of the monthly swing to decrease from maximum to minimum and then to increase back to maximum is 18.61 years.

### 3. Measurement of the Directions of Sunrise and Sunset

The declination $\delta$ of an astronomical object is defined as the complement of the angle between the axis of rotation and the direction of the object, both lines being taken from the centre of the Earth, as in Figure A.3. Obviously, $\delta$ is unaffected by the Earth's rotation; so, if we work from $\delta$, our method will be independent of rotation, as it is required to be. Our aim must be to set up an observational method for measuring $\delta$, and thence to make inferences about sunrise and sunset.

To make observations, convenient local standards of reference are needed. It is natural for an observer on the Earth to choose his own vertical and his own horizontal plane. A standard direction in the horizontal plane is also needed. Any direction – that to some distant tree, for example – will suffice. To the modern mind it is natural to choose

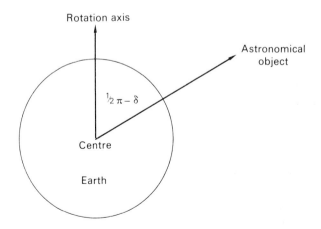

Figure A.3. The definition of the declination $\delta$ of an astronomical object.

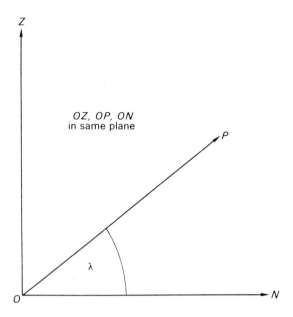

Figure A.4. Measurement of latitude $\lambda$.

the geographical north. Neolithic man may not, and need not, have made this particular choice. However, to permit comparison with texts on spherical astronomy, I will use the geographical north. This is shown as $ON$ in Figure A.4. To obtain $ON$, draw $OP$ parallel to the Earth's rotation axis. The plane $ZOP$ determined by the observer's vertical $OZ$ and by $OP$ intersects the horizontal plane in the line $ON$. The angle $PON$ is the observer's latitude $\lambda$.

In Figure A.5 we have the direction of the Sun referred to this system. The angles $h_s$, $\theta_s$ are the usual elevation and azimuth for this direction, the azimuth being measured east of north. The procedure for finding $h_s$, $\theta_s$ is to take the plane defined by $OZ$ and $OS$, and to let this plane intersect the horizontal plane, say, in the line $OL$. Then $h_s$ is the angle $LOS$ and $\theta_s$ is the angle $NOL$.

The distance of the Sun is very large compared to the radius of the Earth; so the difference between the direction $OS$ for a practical observer on the Earth's surface, and the direction of the Sun for an imaginary observer at the Earth's centre, is very small, and can be neglected. Thus the angle between $OP$ and $OS$ is essentially equal to

$1/2\pi - \delta_s$, $\delta_s$ being the declination of the Sun's centre. The azimuth of $OP$ is zero and the elevation is $\lambda$. With $\lambda$ supposed known, and with $h_s$, $\theta_s$ supposed known for the centre of the Sun, we now have an explicit mathematical problem of finding the angle between two intersecting lines whose elevations and azimuths are known. By a calculation using direction cosines, it is not hard to show that

$$\sin \delta_s = \cos h_s \cos \lambda \cos \theta_s + \sin h_s \sin \lambda. \qquad (1)$$

When the Sun is on the horizon, the angle $h_s$ is small. It is not exactly zero because in practice the horizon does not lie in the observer's horizontal plane. And if we define sunrise to be the moment when the first tip of the Sun appears above the horizon, the centre of the Sun at that time will be depressed below the horizon by half the angular diameter of the solar disc, about 16′.

A further point about $h_s$ must be noted. In the above argument it was assumed that the light from the centre of the Sun reaches the observer along a straight line, the line $OS$ of Figure A.5. But the light

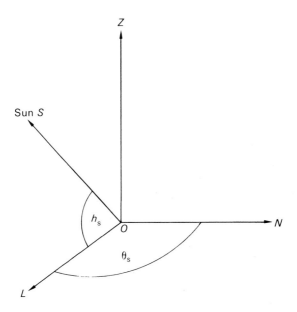

Figure A.5. Measurement of the direction of the Sun in terms of the azimuth $\delta_s$ and the declination $h_s$.

actually follows a curved path because of refraction by the gases of the Earth's atmosphere, and this fact requires us, in order that we may continue to use equation (1), to correct the measured elevation according to Table A.1.

**Table A.1**
*The refraction correction*

| Measured elevation: | $0°$ | $1°$ | $2°$ | $3°$ |
|---|---|---|---|---|
| Correction: | $-35'22''$ | $-24'44''$ | $-18'27''$ | $-14'27''$ |

At Stonehenge, the horizon elevations are remarkably uniform, averaging about 0.6°. This is the measured elevation to be used in assessing the refraction correction of Table A.1. Thus for an horizon elevation of 0.6°, the correction is about $-30'$. Since $h_s$ refers to the centre of the Sun, which is depressed at 'sunrise' by 16′, it follows that $h_s$ is equal to the horizon elevation, say, $h$, *minus* about 46′,

$$h_s = h - 46'. \qquad (2)$$

As stated above, $h_s$ is a small angle, and it is sufficiently accurate to write sin $h_s = h_s$, cos $h_s = 1$, so that (1) becomes

$$\sin \delta_s = \cos \lambda \cos \theta_s + h_s \sin \lambda, \qquad (3)$$

where $h_s$ is expressed in radians.

The next step in analysing the directions of sunrise and sunset lies in relating $\theta_s$ to the position of the Sun in its orbit. Viewed by an observer on the Earth, the Sun appears to move in an orbit that has the same shape as the Earth's orbit has when viewed from the Sun. From the Earth's centre, project every point of the Sun's orbit onto a concentric sphere of very large radius, the celestial sphere. This is done by drawing a line from the Earth's centre to each point of the Sun's orbit, and by extending the line until it intersects the sphere. The result is a great circle on the celestial sphere, the ecliptic. This is the Sun's path 'on the sky.' In a similar way, project the Earth's own equator, to give another great circle, the celestial equator. The two circles intersect at the points ♈ and ♎ of Figure A.6. The point $P$ in this figure is the north pole of the celestial sphere – i.e., the point where the Earth's axis of rotation intersects the celestial sphere. The point $K$ is the pole of the ecliptic.

The Sun appears to move along the ecliptic in the direction of the arrow. The equinoxes occur when the Sun is at ♈ and ♎, the spring

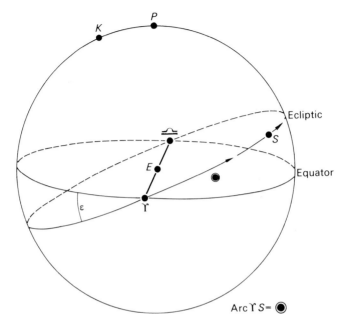

Figure A.6. The ecliptic is the projection of the Sun's orbit on the celestial sphere.

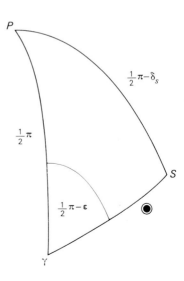

Figure A.7. With $\varepsilon$ and $\odot$ known, the solar declination $\delta$, can be found from this spherical triangle.

equinox at ♈. The angle ☉ from ♈ to S is the longitude of the Sun. When ☉ = 0 we have the spring equinox, ☉ = π/2 is the summer solstice, ☉ = π the autumnal equinox, and ☉ = 3π/2 the winter solstice (π radians equals 180°). By taking great circle arcs from P to ♈ and S, we obtain the spherical triangle shown in Figure A.7. The arc from P to ♈ is π/2 – this follows from the definition of ♈. The arc length ♈ S is ☉, and the angle between P ♈ and S ♈ is π/2 – ε, where ε is the angle between the Earth's equatorial plane and the plane of the ecliptic. At the present day, ε is about 23°27'. At the time of the construction of Stonehenge I, ε was about 24°.

From the mathematical analysis* of the spherical triangle of Figure A.7, it is easy to find $\delta$, in terms of ε and ☉. The result is

$$\sin \delta_{\vee} = \sin \varepsilon \sin \odot. \tag{4}$$

We now have two equations, (3) and (4), both for $\sin \delta_{\vee}$. Equating the righthand sides of these equations, we see that

$$\sin \varepsilon \sin \odot = \cos \lambda \cos \theta_{\vee} + h_{\vee} \sin \lambda. \tag{5}$$

Equation (5) describes the swing of sunrise and sunset at Stonehenge. The results it leads to will be considered later after we arrive at a similar equation for the Moon.

## 4. Measurement of the Directions of Moonrise and Moonset

The relation of the Moon's orbit to the ecliptic is similar to the relation of the Sun's orbit to the celestial equator. This is seen by comparing Figure A.8, for the Moon's orbit, with Figure A.6 for the Sun's orbit. The centre E of the Earth is the centre in both. The plane of the Moon's orbit intersects the plane of the ecliptic in the line N'EN. The points N,N' are the nodes of the lunar orbit, and they play a role in Figure A.8. similar to that played by ♈ and ♎ in Figure A.6. The angle i between the Moon's orbit and the ecliptic is only 5°9', much less than the corresponding angle ε in Figure A.6.

The nodes N, N' are not fixed points. Neither are ♈, ♎, in Figure A.6. Both pairs of points move backward – i.e., clockwise, whereas the Sun and the Moon move counterclockwise. The motion of the line ♎ E ♈ is slow. This 'precession' of the equinoxes has no immediate

---

*The reader wishing to fill in details of this calculation, and of the other calculations indicated below by an asterisk, should consult a textbook on spherical trigonometry.

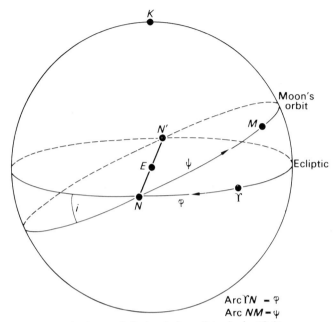

Figure A.8. The angle $\phi$ goes full circuit from 0 to $2\pi$ in 18.61 years. The angle of inclination $i$ of the Moon's orbit to the ecliptic is only 5 9', and has been exaggerated here.

importance for the Stonehenge problem, since it takes about 26,000 years for $\backsimeq E \; \Upsilon$ to make a complete rotation. The rotation of $N'EN$ is much more rapid, however; a complete rotation of this line takes place in 18.61 years, the oscillation period already noted in connection with Figure A.2.

Although the details are more complicated, the procedure now to be followed for the Moon is similar to that already given for the Sun. An equation of the form of (1) holds for the Moon, except that the azimuth $0_s$ of the Sun is replaced by the azimuth $0_m$ of the Moon, and $h_s$ is replaced by $h_m$. The elevation $h_m$ differs from $h_s$, given by (2), because the radius of the Earth cannot be considered small compared with the radius of the Moon's orbit. The correct elevation to be inserted in (1) is for an imaginary observer at the Earth's centre, and this differs from the elevation measured by an observer on the Earth's surface by an amount $a/d$, where $a$ is the radius of the Earth and $d$ is the distance of the Moon. This difference is 57'. Adopting a definition of moonrise and moonset similar to that for sunrise and

117

sunset, we find that the value of $h_m$ to be used in an equation of the form (3), namely,

$$\sin \delta_m = \cos \lambda \cos \theta_m + h_m \sin \lambda, \tag{6}$$

must differ from $h_s$ by 57'. Thus instead of (2), we have

$$h_m = h + 11'. \tag{7}$$

In finding a second equation for $\sin \delta_m$, $\delta_m$ being the declination of the centre of the Moon, we cannot treat the spherical triangle $KNM$ of Figure A.3 in the same way as the triangle $P\Upsilon S$ of Figure A.6, because the declination $\delta_m$ is always defined in relation to the plane of the Earth's equator, not in relation to the plane of the ecliptic. This sets a more ambitious problem in spherical trigonometry. The problem is solved by setting up a system of latitude and longitude on the celestial sphere with $K$ as the pole, and with the great-circle arc $K\Upsilon$ as the zero meridian. The first step is to work out the latitude and longitude of the Moon in this system, taking the arc lengths $\psi = NM$ and $\phi = \Upsilon N$ to be given. These angles obviously determine the position of the Moon uniquely. Next we go back to Figure A.6. Here we could set up two systems of latitude and longitude, one with $P$ as pole and the great-circle arc $P\Upsilon$ as zero meridian, the other with $K$ as pole and the great-circle arc $K\Upsilon$ as zero meridian. The latter is, of course, just the system we have used for the Moon in Figure A.8, whereas the system with $P$ as pole is the one in which $\delta_m$ appears. The essential mathematics of the problem is to relate these two systems.* The outcome is an expression for $\sin \delta_m$ in terms of $\psi$, $\phi$, and of the fixed angles $i$, $\varepsilon$. The result is

$$\sin \delta_m = \sin \psi (\sin i \cos \varepsilon + \cos i \sin \varepsilon \cos \phi)$$
$$- \sin \varepsilon \sin \phi \cos \psi, \tag{8}$$

which is the analogue of equation (4). We can now proceed to equate the right-hand sides of (6) and (8), to give

$$\cos \theta_m \cos \lambda = \sin \psi (\sin i \cos \varepsilon + \cos i \sin \varepsilon \cos \phi)$$
$$- \sin \varepsilon \sin \phi \cos \psi - h_m \sin \lambda. \tag{9}$$

Equation (9) determines the azimuth $\theta_m$ of moonrise, since the first tip of the rising moon has just the same azimuth as the centre of the Moon. The angle $\psi$ is the analogue of the solar longitude $\odot$. The angle

---

*See a text on spherical astronomy.

118

$\odot$ determines the seasons of the year, and $\psi$ determines the phases of the Moon. Thus $\odot$ increases from 0 to $2\pi$ in a year, whereas $\psi$ increases from 0 to $2\pi$ in a lunar month of 27.322 days. The complication in (9) arises from the angle $\phi$. Looking at Figure A.8, we see that $\phi$ determines the position of $N$ in relation to $\Upsilon$. This angle increases from 0 to $2\pi$ in 18.61 years. During any particular month, $\phi$ can be considered to be essentially constant, and the change of $\psi$ from 0 to $2\pi$ produces a swing back and forth in the azimuth of moonrise. However, as $\phi$ changes from month to month, the angle of swing of moonrise changes. The largest swing occurs when $\phi=0$, and the smallest swing occurs when $\phi=\pi$.

## 5. The Situation at Stonehenge

The alignments of Figure A.1 for sunrise and sunset are found by putting $\sin \odot = \pm 1$ in equation (5). Putting $\sin \odot = 1$, and using the fact that the elevation $h_s$ is a small angle, one can without undue difficulty show that, to sufficient accuracy,

$$\theta_s = \cos^{-1}\left(\frac{\sin \varepsilon}{\cos \lambda}\right) + \frac{h_s \sin \lambda}{\sqrt{\cos^2 \lambda - \sin^2 \varepsilon}}. \tag{10}$$

With $\lambda = 51°11'$ for the latitude of Stonehenge, and with $\varepsilon = 24°$ for the epoch of construction of Stonehenge I, equation (10) gives

$$\theta_s = 49.54° + 1.63\, h_s \tag{11}$$

for the azimuth of sunrise on midsummer day. When $\sin \odot = -1$, the corresponding result for the azimuth of sunrise on midwinter day is given by

$$\theta_s = \pi - \cos^{-1}\left(\frac{\sin \varepsilon}{\cos \lambda}\right) + \frac{h_s \sin \lambda}{\sqrt{\cos^2 \lambda - \sin^2 \varepsilon}}$$
$$= 130.46° + 1.63\, h_s. \tag{12}$$

The most northerly rising of the Moon is given by putting $\phi=0$, $\psi=\pi/2$ in (9), and is

$$\theta_m = \cos^{-1}\left(\frac{\sin(i+\varepsilon)}{\cos \lambda}\right) + \frac{h_m \sin \lambda}{\sqrt{\cos^2 \lambda - \sin^2(i+\varepsilon)}}$$
$$= 39.01° + 1.97\, h_m; \tag{13}$$

the most southerly rising of the Moon is given by putting $\phi=0$, $\psi=3\pi/2$ in (9), and is

$$\theta_m=\pi-\cos^{-1}\left(\frac{\sin(i+\varepsilon)}{\cos\lambda}\right)+\frac{h_m\sin\lambda}{\sqrt{\cos^2\lambda-\sin^2(i+\varepsilon)}}$$

$$=140.99°+1.97\,h_m. \tag{14}$$

The inner dotted lines of moonrise in Figure A.2 are given by putting $\phi=\pi$, $\psi=\pi/2$ and $\phi=\pi$, $\psi=3\pi/2$ in (9), with the results

$$\theta_m=\cos^{-1}\left(\frac{\sin(\varepsilon-i)}{\cos\lambda}\right)+\frac{h_m\sin\lambda}{\sqrt{\cos^2\lambda-\sin^2(\varepsilon-i)}}$$

$$=58.97°+1.45h_m, \tag{15}$$

and

$$\theta_m=\pi-\cos^{-1}\left(\frac{\sin(\varepsilon-i)}{\cos\lambda}\right)+\frac{h_m\sin\lambda}{\sqrt{\cos^2\lambda-\sin^2(\varepsilon-i)}}$$

$$=121.03°+1.45h_m, \tag{16}$$

respectively.

The corresponding azimuthal values for sunset and moonset are given by subtracting the angles (11) to (16) from $360°$.

With the above definitions of sunrise and sunset, and of moonrise and moonset, $h_s$ and $h_m$ are given in terms of the horizon elevation $h$ by equations (2) and (7). Notice that the code numbers $\pm24$, $\pm29$, and $\pm19$ used in Chapter 2 are nothing but the values in degrees of $\pm\varepsilon$, $\pm(\varepsilon+i)$, $\pm(\varepsilon-i)$; so the alignments associated with these code designations are simply those given in (11) to (16). In terms of the horizon elevation $h$, we thus have the azimuthal alignments set out in Table A.2. The calculated angles used in Chapter 4 were taken from this table.

**Table A.2**
*Azimuthal alignments*

| Number | Sunrise | Sunset |
|---|---|---|
| $+24$ | $48.3+1.63h$ | $311.7-1.63h$ |
| $-24$ | $129.2+1.63h$ | $230.8-1.63h$ |
| $+29$ | $39.4+1.97h$ | $320.6-1.97h$ |
| $-29$ | $141.4+1.97h$ | $218.6-1.97h$ |
| $+19$ | $59.2+1.45h$ | $300.8-1.45h$ |
| $-19$ | $121.3+1.45h$ | $238.7-1.45h$ |

## 6. Offsets from the Astronomical Alignments

Starting with the simpler solar case, we can, with enough accuracy for our purposes here, omit the elevation term from equation (5). Suppose a sighting line to be directed by a small angle $\chi$ inside the most northerly annual swing of the Sun. How long before midsummer day will the Sun cross this line?

We relate the direction of sunrise $\theta_s$ to the longitude $\odot$ by the equation

$$\cos\theta_s = \frac{\sin\varepsilon\sin\odot}{\cos\lambda}. \tag{17}$$

The Sun rises on the line in question when $\theta_s = \cos^{-1}(\sin\varepsilon/\cos\lambda) + \chi$. Writing $\odot = \pi/2 - \Delta$, we can easily show that equation (17) leads to

$$\Delta^2 = \frac{2\sqrt{\cos^2\lambda - \sin^2\varepsilon}}{\sin\varepsilon}\chi, \tag{18}$$

with $\Delta$, $\chi$ in radians. Putting $\lambda = 51°11'$, $\varepsilon = 24°$, and measuring $\Delta, \chi$ in degrees, one obtains

$$\Delta^2 = 134.37\chi. \tag{19}$$

For $\chi = 1°$, equation (19) gives $\Delta = 11.59°$; and for $\chi = 2°$, $\Delta = 16.39°$. Since the Sun moves in longitude by about $1°$ per day, the Sun will cross sighting lines at $\chi = 1°$, and at $\chi = 2°$, about 12 days and 16 days, respectively, before midsummer. For a stone that produces a blank arc on the horizon between $\chi = 1°$ and $\chi = 2°$, the blank sequence in which the tip of the rising sun is not seen would last for four days.

Turning now to the Moon, let us first find the most northerly rising that can occur when the position of the node $N$ is determined by a general value of $\phi$, in the manner of Figure A.8. In terms of the lunar longitude $\psi$, the direction of $\theta_m$ of moonrise is given by equation (9). Again omitting the elevation term, we can write

$$\cos\theta_m\cos\lambda = X\sin\psi + Y\cos\psi, \tag{20}$$

where

$$X = \sin i\cos\varepsilon + \cos i\sin\varepsilon\cos\phi,$$
$$Y = -\sin\varepsilon\sin\phi. \tag{21}$$

In a particular lunar month, $\psi$ goes full circle, but $\phi$ hardly changes. If we make the slight approximation of keeping $\phi$ strictly constant, the minimum of $\theta_m$ with respect to $\psi$ occurs for

$$\tan \psi = X/Y, \tag{22}$$

and is given by $\cos^{-1}(\sqrt{X^2 + Y^2}/\cos \lambda)$. This is the most northerly swing of moonrise *at the value of $\phi$ in question*. For $\phi$ small, after some reduction, we have

$$\theta_m = \cos^{-1}\left[\frac{\sin(i+\varepsilon)}{\cos \lambda}\right] + \frac{1}{2}\phi^2 \frac{\sin i \sin \varepsilon \cos(i+\varepsilon)}{\sin(i+\varepsilon)\sqrt{\cos^2 \lambda - \sin^2(i+\varepsilon)}}. \tag{23}$$

The first term on the right-hand side gives the most northerly rising of the Moon for the whole 18.61-year cycle, and the second term on the right-hand side is the amount by which the most northerly swing of moonrise, in the particular month determined by the value of $\phi$, falls short of the extreme for the whole cycle. Writing

$$\theta_m = \cos^{-1}\left[\sin(i+\varepsilon)/\cos \lambda\right] + \chi,$$

with $\chi$ again the offset, we thus have

$$\chi = \frac{1}{2}\phi^2 \frac{\sin i \sin \varepsilon \cos(i+\varepsilon)}{\sin(i+\varepsilon)\sqrt{\cos^2 \lambda - \sin^2(i+\varepsilon)}}. \tag{24}$$

Inserting $i = 5°9'$, $\lambda = 51°11'$, $\varepsilon = 24°$ in (24), and measuring $\phi$, $\chi$, in degrees, we therefore get

$$|\phi| = 26.3\chi^{1/2}, \tag{25}$$

which gives the pairs of related values shown in Table A.3. In this table, the angle $\phi$ is considered to be negative before, and positive after, the most northerly rising of the Moon in the 18.61-year cycle, and $|\phi|$ is the positive magnitude of $\phi$.

**Table A.3**
*Values of $\chi$ and $|\phi|$*

| $\chi$ | 0.5 | 0.6 | 0.7 | 0.8 | 0.9 | 1.0 | 1.25 | 1.5 | 1.75 |
|---|---|---|---|---|---|---|---|---|---|
| $|\phi|$ | 18.6 | 20.4 | 22.0 | 23.5 | 24.9 | 26.3 | 29.4 | 32.2 | 34.8 | 37.2 |

The angle $\phi$ changes by $360 \times 27.32/365.25 \times 18.61 = 1.45°$ in a lunar month of 27.32 days, equivalent to a change of about $0.1°$ in $\chi$. It follows that a wooden post which establishes a blank arc of about 0.2° in $\chi$ would obscure in two consecutive months the most northerly swing of the rising moon, a result stated in Chapter 4.

Since the change of $\phi$ in a year is $360/18.61 = 19.3°$, more than a year is needed to go from $\chi = 1°$ to $\chi = 0$, a result also quoted in Chapter 4. Indeed, almost two years are required to go from $\chi = 2°$ to $\chi = 0$.

Similar results apply for offsets from the other three solar alignments and from the other seven lunar alignments.


## 7. An Error Analysis

The problem for an observer using the sighting lines of Stonehenge I was to set markers $S,N,N'$, on the Aubrey Circle to represent within a tolerance of about $\pm 1°$ the true instantaneous positions of the Sun and the lunar nodes on the ecliptic. Observations are of necessity confined to the azimuthal directions of the rising Sun and Moon. Yet small errors in estimating these angles translate into much larger errors in the angles $\odot, \phi$ which determine the true positions of the Sun and the lunar node.

Consider, first, the error due to the positioning of the observer. Inadvertent variations of two inches in the observer's eye position, when taken over a sighting line of 200 feet, give angular variations of 0.048°. Such variations, in effect, change the offset angle $\chi$ by 0.048°. From (19) and (25), the changes (in degrees) in $\odot$ and $\phi$ due to such a change in $\chi$ are $0.28/\sqrt{\chi}$ and $0.63/\sqrt{\chi}$, respectively. It follows that variations of eye position are acceptable, provided the offset angle $\chi$ is not much less than 0.5°, and provided that the sighting lines are at least 200 feet in length.

Let us come, second, to a more subtle form of error. From (20), we saw that the smallest value of the azimuthal angle $\theta_m$ of the rising moon that can occur at a particular value of $\phi$ is given by

$$\tan \psi = \frac{X}{Y}, \theta_m = \cos^{-1}\left(\frac{\sqrt{X^2 + Y^2}}{\cos \lambda}\right), \tag{26}$$

$X,Y$ being given by (21). The first of equations (26) gives an explicit value for the lunar longitude $\psi$, which gives an explicit position of the Moon in its orbit around the Earth. But will the Moon happen to be rising just at the moment when $\psi$ attains this particular value?

## On Stonehenge

There will certainly be some place on the Earth where it is rising, but Stonehenge will not in general be that place. Indeed, moonrise will usually occur at Stonehenge when $\psi$ differs from $\tan^{-1}(X/Y)$ by several degrees. This deviation of $\psi$ is not fixed – it varies from month to month, depending on where the Moon is in its orbit in relation to a point on the surface of the spinning Earth. The largest deviation occurs when moonrise at Stonehenge precedes (or follows) by a half-day the moment when $\psi = \tan^{-1}(X/Y)$. In this situation $\psi$ at the moment of moonrise differs from $\tan^{-1}(X/Y)$ by some $6°$ to $7°$. On the average, however, the deviation is about $3°$.

Write $\delta\psi$ for the deviation of $\psi$ from $\tan^{-1}(X/Y)$ at the moment of moonrise at Stonehenge. What is the effect of these deviations on the azimuthal angle $\theta_m$? From (20), it is not hard to show that (23) is changed to

$$\theta_m = \cos^{-1}\left[\frac{\sin(i+\varepsilon)}{\cos\lambda}\right] + \frac{1}{2}\phi^2\frac{\sin i \sin\varepsilon \cos(i+\varepsilon)}{\sin(i+\varepsilon)\sqrt{\cos^2\lambda - \sin^2(i+\varepsilon)}}$$
$$+ \frac{1}{2}\delta\psi^2\frac{\sin(i+\varepsilon)}{\sqrt{\cos^2\lambda - \sin^2(i+\varepsilon)}} \qquad (27)$$

Neglecting a higher-order term, we can rewrite (27) in the form,

$$\theta_m = \cos^{-1}\left[\frac{\sin(i+\varepsilon)}{\cos\lambda}\right]$$
$$+ \frac{1}{2}(\phi+\delta\phi)^2\frac{\sin i \sin\varepsilon \cos(i+\varepsilon)}{\sin(i+\varepsilon)\sqrt{\cos^2\lambda - \sin^2(i+\varepsilon)}}, \qquad (28)$$

where

$$\delta\phi = \frac{1}{2}\frac{\delta\psi^2}{\phi}\frac{\sin^2(i+\varepsilon)}{\sin i \sin\varepsilon \cos(i+\varepsilon)}. \qquad (29)$$

Hence the effect of the deviation $\delta\psi$ is like introducing an error into $\phi$ of the amount given by the right-hand side of (29). Inserting $i = 5°9'$, $\varepsilon = 24°$, we get

$$\delta\phi = 3.72°\frac{\delta\psi^2}{\phi}. \qquad (30)$$

with all angles in radians.

The angle $\phi$ is related to the offset $\chi$ of the sighting line by (25), $\phi$ being negative before, and positive after, the most northerly rising

124

of the Moon is the 18.61-year cycle. Averaging $\delta\phi$ before and after the most northerly rising, from (25) and (30) we get

$$\frac{0.0707}{\sqrt{\chi}} \ (\delta\psi_a^2 - \delta\psi_b^2), \tag{31}$$

where $\delta\psi_a$, $\delta\psi_b$ are the deviations of $\psi$ from $\tan^{-1}(X/Y)$ in the particular months when the sighting line is crossed, after and before the most northerly rising of the Moon in the 18.61-year circle. The error (31) is as likely to be negative as positive. With $\delta\psi_a$, $\delta\psi_b$ taken to lie with uniform probability anywhere in the range from $-6°$ to $+6°$, it can be shown that the average value of the magnitude of (31) is $0.85/\sqrt{\chi}$ degrees, which again does not much exceed $1°$ unless the offset angle is appreciably less than about $0.5°$.

In practice, we would be concerned not with the crossing of a single line, but with the blank arc determined by the width of a stone or post. The blank-sequence technique itself introduces an error, additional to (31). We have to find either the month when the most northerly rising of the Moon first becomes obscured by the blank arc, or the month in which the most northerly rising emerges out of the blank arc. Thus the blank sequence method has a built-in discreteness – time jumps that are measured in lunar months, not in days or half-days. Since $\phi$ changes by $1.45°$ per lunar month, the resulting error in $\phi$ can be anywhere from zero to $1.45°$ with uniform probability, the average error being $1/2 \times 1.45 = 0.72°$. This average error in the setting of the nodal marker $N$ is inherent in the blank-sequence method. However, provided the offset angle $\chi$ is not taken too small, the several sources of error in the setting of the nodal marker all contribute averages less than $1°$. Taken together, the combined error should still be less than $2°$, which can be considered an acceptable tolerance.

In giving the above error analysis, I am not seeking to imply that these considerations were known to neolithic man, for it can scarcely be supposed that neolithic man knew either the detailed astronomical facts on which the equations are based, or the mathematical methods for handling the equations. But the conclusions of the analysis could have been discovered by actual trial. For example, an important conclusion of the analysis is that accuracy becomes impaired when the offset angle $\chi$ is too small. This same conclusion could be reached by trial observations at the site, by examining the results obtained from sighting lines constructed with different offsets, ranging from less than $0.5°$ to offsets of several degrees, just as Table 4.1 shows was done.

Perhaps the most remarkable aspect of Stonehenge is to find that it all works out successfully. We started from the simple coincidental associations of Chapter 2. The fact that no serious difficulty is encountered as one delves deeper into the procedures for operating Stonehenge as an eclipse predictor, provides, in my view, a very strong argument for thinking that Stonehenge was indeed used for this purpose.

## Bibliography

W. M. Smart, *Spherical Astronomy*. Cambridge University Press, 1956.

# *Index*

# Index